QUANTITATIVE PAPER

AND

THIN-LAYER

CHROMATOGRAPHY

QUANTITATIVE PAPER AND THIN-LAYER CHROMATOGRAPHY

EDITED BY

E. J. SHELLARD

Department of Pharmacy
Chelsea College
London, England

1968

ACADEMIC PRESS
LONDON AND NEW YORK

ACADEMIC PRESS INC. (LONDON) LTD
24-28 Oval Road
London NW1

U.S. Edition published by
ACADEMIC PRESS INC.
111 Fifth Avenue
New York, New York 10003

Library of Congress Catalogue Card Number: 68-9114
SBN: 12-639750-3

Printed in Great Britain by
The Whitefriars Press Ltd
Tonbridge, Kent

List of Contributors

W. E. COURT, *School of Pharmacy, College of Technology, Liverpool, England*

J. W. FAIRBAIRN, *Department of Pharmacognosy, School of Pharmacy, University of London, London, England*

G. FRANGLEN, *Department of Chemical Pathology, St. George's Hospital Medical School, University of London, London, England*

G. W. GOODMAN, *Research Laboratories, B.P. Chemicals (U.K.) Ltd., Epsom, Surrey, England*

D. E. JÄNCHEN, *Camag, Muttenz, Switzerland*

C. A. JOHNSON, *Scientific Director, British Pharmacopœia Commission, London, England*

H. JORK, *Department of Pharmacognosy and Analytical Phytochemistry, University of the Saarland, Saarbrücken, German Federal Republic*

A. E. LOWE, *Biochemistry Department, Imperial College, University of London, London, England*

B. R. PULLAN, *Department of Medical Physics, University of Manchester, England*

E. J. SHELLARD, *Department of Pharmacognosy, Chelsea College of Science and Technology, University of London, London, England*

B. A. WOOD, *Department of Drug Metabolism, Pfizer Limited, Sandwich, Kent, England*

Preface

It was inevitable that, with the development of paper and thin-layer chromatographic techniques for the separation of very small quantities of mixtures into their individual components, some attempts should be made to determine the amounts of the constituents present. Early attempts which involved the measurement of the spot dimensions were made by Fisher, Parsons and Holmes in 1949, and in 1962 Purdy and Truter introduced methods based on a mathematical relationship between the spot area and the amount of substance present, methods which are still very useful. However, the extensive use of physical chemical methods of evaluation has spread to paper and thin-layer chromatographic separations.

Two general methods are used: the determination of the substance after its elution from the paper or adsorbent, and the determination of the substance directly on the paper or thin-layer.

Results obtained by the application of the various analytical techniques to one or another of these two methods are being reported by an increasing number of workers and it seemed desirable that a forum should be arranged for a general discussion on the problems associated with quantitative paper and thin-layer chromatography. The Department of Pharmaceutical Sciences (Pharmacognosy Committee) of the Pharmaceutical Society, decided to organize a Symposium and later was joined by the Thin-Layer Chromatography Group of the Society for Analytical Chemistry. Justification for the organization of the Symposium was shown by the attendance of nearly 400 people.

This publication consists of the contributions made by the invited speakers at the Symposium held at the Chelsea College of Science and Technology on January 3rd and 4th, 1968, though the order differs slightly from that in which they were presented. Although a number of interesting points were raised during the discussion it has been decided not to include them in this publication. There is no doubt, that quantitative paper and thin-layer chromatography has great potential but much more work needs to be done—and possibly more symposia organized—before reproducibility and precision of results can be guaranteed. Nevertheless it was felt that analysts, biochemists, pharmacognosists, pharmacologists, phyto-chemists and others who use chromatographic techniques would benefit from reading a full account of the Symposium papers.

The Editor would like to thank all the contributors for readily producing copies of their papers and thus making his task relatively easy, and the publishers for their enthusiasm and assistance in ensuring that the time between the date of the Symposium and that of publication has been kept to a minimum.

June 1968   E. J. SHELLARD

Contents

Chapter 1

Factors Involved in Producing Uniform Spots

J. W. FAIRBAIRN

Department of Pharmacognosy, The School of Pharmacy,
University of London, London, England

Quantitative layer chromatography is attractive because of its simplicity and in principle it should be accurate since, in identical circumstances, the quantity of substance in the final spot will always be in a fixed ratio to the quantity applied in the initial or starting spot. The pattern of distribution in the final spot should also be constant so that the ultimate measurement (spot area, densitometer reading or optical density, after elution) should accurately represent the amount of substance originally applied. In practice, however, it is extremely difficult to assure identical conditions for every chromatographic run and the object of this paper is to discuss the factors which cause variations in the final spots and to suggest means of overcoming them.

Several workers have studied this problem in detail, notably Bush (1961, 1963). Experience in our laboratories confirms the importance of equilibrating the paper overnight, maintaining a constant temperature during the run, in an undisturbed tank containing an "initial solvent volume" of 1%. The initial spots should be of constant area and as small as possible (Fairbairn and Suwal, 1959; Morrison and Orr, 1966). Drying and spraying should also be carried out in controlled conditions (Bush, 1961). To meet certain conditions which are not under the control of the worker, such as variations within and between sheets of paper, some workers recommend the use of standard and control solutions for every determination on the basis of the "four-point bioassay" technique (Fairbairn and Suwal, 1959; Fisher, *et al.*, 1949; Ohtsu and Mizuno, 1952).

The success of the precautions taken by workers will be reflected in the reproducibility of the method. Unfortunately not all workers test their reproducibility by proper statistical analysis of an adequate number of replicates. One well-recognized statistic is the coefficient of variation (standard deviation of individual results expressed as the percentage of the mean). If the individual results form a normal distribution then 95% ($P = 0.95$) of the results will fall within a range of approximately twice the coefficient of variation. When insufficient replicates have been made it may not be possible to check whether the distribution is normal or to estimate the true standard deviation; the variation may therefore be three times the coefficient of variation, or even more. A coefficient of variation of 5% will therefore indicate that individual results will vary ($P = 0.95$) ± 10–15% of the mean. Obviously if the figures quoted by workers are themselves the *means* of several results, the variation of these figures will be less than that for individual results. This "standard error of the mean" can be obtained by dividing the variation of individual results by the square root of the number of results used in arriving at the mean. Some examples of published variations for quantitative chromatographic methods are: Fairbairn and Wassel (1963), coefficient of variation 6–6.8%; McEvoy-Bowe and Lugg (1961), coefficient of variation (calculated from their standard error of the mean of triplicate assays) 7.7% Römisch (1961), coefficient of variation (estimated from his figures (8–16%; Genest and Farmilo (1959) $\pm 4\%$. Bush (1963) quotes maximum variations of $\pm 5\%$ or $\pm 7\%$ for experienced workers; presumably these would be equivalent to coefficients of variation of about 3%. In a review of quantitative thin-layer methods, Pataki (1967) gives errors ranging from $\pm 3.5\%$ to ± 10–15%. Recent attempts in the author's laboratory to estimate opium alkaloids by densitometry of the final spots still led to coefficients of variation of ± 5.2 to $\pm 6.7\%$, even when all the precautions mentioned previously were observed. We, therefore, decided to make a systematic study of possible sources of error. In order to exhibit these errors as sensitively as possible all the statistics were calculated on the basis of individual results, rogue results were not excluded nor were means of groups dealt with in order to cover up the odd figures. In practice, of course, most workers ignore obvious rogue results and where there is unavoidable variation use means of several results. However for satisfactory statistical analysis all individual data must be used; the resulting analysis then forms a sound basis for the worker to decide how many replicates would be necessary to reduce errors to a convenient minimum.

Errors can arise during the following three phases of chromatography, (a) production of the initial spots, (b) translocation of substance from the initial to the final spot and (c) treatment of the final spot and subsequent measurement. The first two sets of factors were studied using radioactive substances for convenience in analysis and as will be seen a hitherto largely unsuspected source of error was brought to light and its reduction has led to a significant improvement in the uniformity of the final spot. The information gained was then applied to the study of the errors likely to arise in measuring the quantities in the final spot; for this latter purpose densitometry was used.

A. PRODUCTION OF THE INITIAL SPOTS

Little attention seems to have been directed to the problem of whether the measuring instruments used actually do deliver replicates of constant volume when forming the initial spots. Volumes of 5, 2 or even 1 μl. are frequently used and, clearly, any errors involved here would vitiate all attempts to produce identical conditions in subsequent stages. Three common measuring instruments are the Agla syringe, the Hamilton syringe and the Drummond Microcap micropipette (a calibrated capillary tube surmounted by a small rubber teat). Obviously a certain amount of experience is necessary to obtain maximum accuracy; accordingly experienced workers were invited to measure out replicate volumes of solution exactly as they normally did in their quantitative work, using their favourite syringe or pipette. Since a large number of analyses were involved radioactive chemicals were used so that quantities could be rapidly measured as radioactivities in an automatic scintillation counter. The errors involved in using the counter were approximately $\pm 1 \cdot 5\%$ (coefficients of variation, as already defined, ranged from $0 \cdot 45\%$ to $0 \cdot 78\%$). Full details are given in the paper by Fairbairn and Relph (1968a).

Radioactive glucose, tyrosine and morphine were used both in paper and thin layer. Each worker was asked to deliver 10–16 replicates (a) direct into glass vials containing the phosphor for radiactive counting, and (b) on to paper or thin layer plates. Exactly similar areas of paper or absorbent containing each spot from (b) were transferred to phosphor and their radioactivities determined. The results, given in Table 1, are quite striking as they reveal errors ranging from ± 6 to $\pm 20\%$ or more for individual measurements. They therefore indicate a major unsuspected source of error in quantitative layer chromatography since most workers have assumed they can measure their initial volumes accurately and have concentrated on sources of error likely to arise at later stages.

1. SOURCES OF ERROR IN VOLUME MEASUREMENTS

A careful examination of the process of delivering the required volume indicated two sources of variation.

a. Creep back

The delivery of a number of drops from the needle of an Agla syringe by free fall was observed with a lens and it was noted that from time to time an accumulating drop suddenly slipped up slightly from the point and when the drop finally fell on to the paper a definite proportion remained on the stem. This "creep back" effect was cumulative and sometimes a sizeable volume remained on the stem for some time; then quite unpredictably it would disappear with a succeeding drop. The creep back effect varied with the solvent used and was particularly noticeable with methanol; occasionally after the delivery of many drops of a methanolic solution of morphine, crystals of the latter substance were seen to have formed a tide mark on the stem of the needle. At times an exceptionally high tide of

TABLE 1

Errors due to measurement of small volumes of solution by experienced workers, expressed as coefficient of variation (s.d. of individual results calculated as a percentage of the mean)

Worker	Solute	Volume of solution measured (μl.)	Coefficient of variation	
			Direct into phosphor (%)	Via adsorbent (%)
A	Morphine-2-T	10 (Agla)	7·8	7·3 (Paper)
B	Morphine-2-T	10 (Agla)	5·8	10·1 (Paper)
C	Morphine-2-T	2 (Micropipette)	6·6	3·3 (Paper)
C	D-glucose-^{14}C(U)	2 (Micropipette)	4·7	
		5 (Micropipette)	2·5	
D	D-glucose-^{14}C(U)	1 (Hamilton)	11·1	8·2 (Paper)
			10·4	7·8 (Paper)
		1 (Hamilton)		8·8 (TLC)
				9·8 (TLC)
B	D-glucose-^{14}C(U)	5 (Agla)	5·5	
E	D-glucose-^{14}C(U)	5 (Agla)	3·4	3·2 (Paper)
				6·0 (TLC)
F	L-tyrosine-^{14}C(U)	5 (Agla)	4·5	9·1 (Paper)
		10 (Agla)	10·9	11·9 (Paper)
B	L-tyrosine-^{14}C(U)	5 (Agla)	7·9	
G	L-tyrosine-^{14}C(U)	5 (Agla)	4·7	

creep back would reach the crystals, redissolve them, and wash them into a succeeding drop. Accordingly some of the needles used in the experiments recorded in Table 1 were washed after each series of radioactive solutions had been delivered. In several cases the "remainder" on the needle represented a significant proportion of the total radioactive substance measured, e.g. for morphine-2-T (Worker B) after delivering 16 volumes the remainder represented 9·6% of the total delivered. Worker E, on the other hand, avoided creep back by using a hooked needle previously dipped in silicone; there was no detectable remainder after his series of deliveries.

b. Capillation

A second source of error arises from the fact that the measured drop does not always fall freely from the end of the needle as its weight may be insufficient to overcome surface tension effects. To assist this most workers touch the drop on to the surface of the paper or adsorbent or against the glass vial. We found that this process withdrew fluid from the lumen of the needle by capillarity and sometimes it required the delivery of 0·6–0·8 μl. of solution from the barrel of the syringe into the needle before the succeeding drop made its appearance. The amount withdrawn varied according to the bore of the needle, time of contact and state of absorbency of the absorbent material, which decreases as more liquid is added to it.

It is interesting to note that with a micropipette the degree of creep back is greatly reduced since only one delivery is made and capillation is probably taken into account when calibrating. These facts may explain why the errors of Worker C (Table 1) were somewhat lower than those of the workers using syringes. As the latter are obviously preferred for replicates or for measuring varying volumes an attempt was made to overcome the two sources of error by (a) siliconing the needles, (b) using different bore thicknesses, and (c) filing the tapered ends till they were at right angles to the long axis, but none of these methods were entirely successful. Success was finally obtained by means of a machine designed by Bridger and Relph (1967) which automatically delivers small volumes by rapid ejection or throwing (the Chromaplot, Fig. 1). Creep back was prevented and since even quite small drops could be forced on to the paper or adsorbent without touching the surface, capillation was also prevented. This twofold advantage was

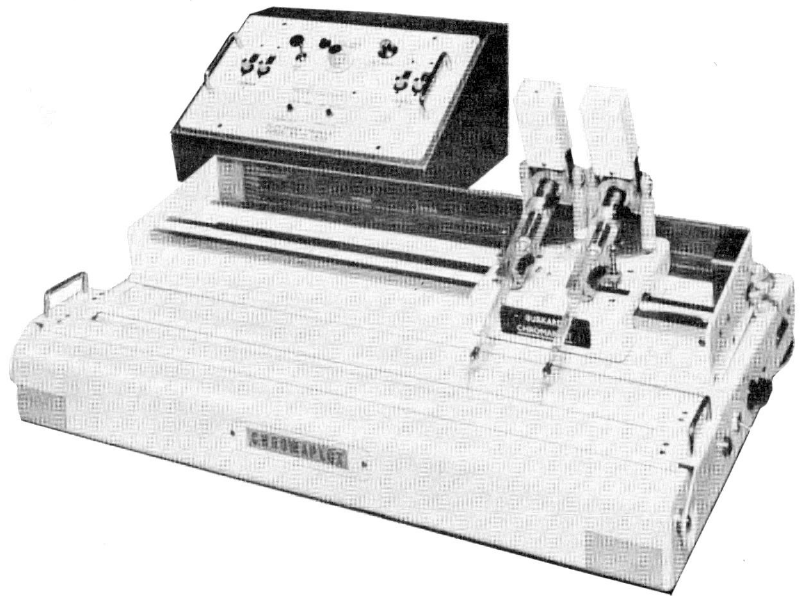

FIG. 1. Photograph of the Chromaplot, an instrument designed for accurate delivery of small replicate volumes by forcible ejection. Two syringes are available, one for the standard solution and the other for the test solution. The printed circuit key at the rear, which can be interchanged with others as required, programs automatic delivery of spots, short bands or streaks. The instrument is produced commercially by Burkard, Rickmansworth, Herts, U.K.

demonstrated by producing a series of drops from the machine by forcible ejection in the normal way; the coefficient of variation was found to be $\pm 2 \cdot 5\%$. A second series was produced from the machine in identical circumstances except that the needle was brought sufficiently near to the paper for the drops to be drawn off by capillation, that is, to fall rather than be thrown. In these conditions the coefficient of variation rose to $\pm 7\%$. Creep back had also occurred as the remainder left on the needle after this series represented $3 \cdot 27\%$ of the total quantity delivered. When the normal throwing procedure was used the remainder was only $0 \cdot 32\%$.

In Table 2 some results obtained with the new machine are shown and it is obvious that the errors are considerably less than those for hand delivered drops (Table 1). The results are also more consistent, tedious hand spotting with likelihood of fatigue is avoided and the resulting spots are small and of constant area since equal volumes ($0 \cdot 2$–$0 \cdot 4$ $\mu l.$) are ejected each time and the circle of solution allowed to dry before the next drop is ejected.

TABLE 2

Errors due to measurement of small volumes of solution
using the machine referred to in the text

Worker	Solute	Volume of solution measured ($\mu l.$)	Coefficient of variation	
			Direct into phosphor (%)	Via adsorbent (%)
B	Morphine-2-T	2	2·4	1·8 (Paper)
B	Morphine-2-T	5	2·2	2·5 (Paper)
B	Morphine-2-T	9·6	1·6	2·1 (Paper)
B	D-glucose-^{14}C(U)	5	2·0	2·7 (Paper)
B	L-tyrosine-^{14}C(U)	5	1·5	1·5 (Paper)

B. Translocation of the Substances from the Initial to the Final Spot

Three main sources of error are likely to occur during this phase; (a) lateral diffusion of the solutes during the run, (b) variations in the structure of different sheets or layers, and (c) variations from tank to tank. The latter was eliminated by always using the same tank in identical conditions, that is, in a constant temperature water bath and a critical solvent volume of 1% as recommended by Block *et al.* (1958). The other two groups of factors were investigated using radioactive glucose on paper chromatograms for convenience.

1. LATERAL DIFFUSION EFFECTS

These will operate as soon as the mobile phase has passed the starting line, since there will always be some solute molecules dissolved in the continuous sheet of flowing solvent. The extent of this lateral diffusion will depend on the relative rate of diffusion of the solute and that of the ascending, or descending, mobile phase. Evidently some does take place as the final spot is normally of greater diameter than the initial one. Errors would obviously arise if varying amounts of solute diffused out of the normal pathway during the run. This export of molecules, however, may be compensated for by an exactly equal import from neighbouring running spots. However, the export-import balance may vary considerably owing to variation in the distance between and in the amount of substance in neighbouring spots and whether the pathway is against the edge of the paper or in the middle.

Results obtained by Fairbairn and Relph (1968*b*) showed that no significant errors arose when similar quantities of glucose were chromatographed even though the distance between the spots varied from 4 to 30 cm or when a single spot was run on a sheet. In all instances the proportion of glucose in each final spot was substantially the same, with a coefficient of variation (1·26%) no greater than that (2·06%) involved in producing the initial spots using the machine designed to this purpose. However when a more stringent test was applied by running dissimilar quantities (20 μg and 200 μg) alternately so that each glucose poor pathway was flanked on either side by glucose-rich pathways, and vice versa, there was some evidence that slight excesses of glucose had diffused into the glucose poor spots. The results are shown in Table 3 and though there is no statistically significant difference between the means for the 20 μg and 200 μg spots it is clear that the individual values for the 20 μg spots are consistently higher than those for the 200 μg spots. Although the amount of lateral diffusion is still small it would be advisable to avoid it by not running two dissimilar spots side by side.

2. VARIATION BETWEEN SHEETS OR LAYERS

Although manufacturers take considerable care in producing chromatographic sheets it is obvious that in such complex structures variations between sheets will occur. This may be more pronounced in TLC where most workers use hand spreaders when preparing their plates. The results as already reported, however, indicate that little variation results during the paper chromatography of similar amounts of substance. However, the problem arising in quantitative work is whether the results obtained from a known quantity of standard can be safely extrapolated to that obtained from an unknown, and almost certainly different, quantity of test. The question arises whether the slope of the regression line relating measurement of the final spot with quantities applied to the initial spots is consistent within one sheet and if so, whether the slope is identical with that of other sheets run in

TABLE 3

Variations in radioactivities of final spots derived from
(a) initial spots of low and high concentrations placed alternately and
(b) initial spots run singly

Initial spots (in order) (μg)	dpm/50 μg $\times 10^{-2}$	Final spots Means	Coefficient of variation (%)
(1) 20	96·87		
(2) 200	96·41		
(3) 20	94·90		
(4) 200	93·51	4 $\times$ 20 μg = 97·46	2·78
(5) 20	101·29	4 $\times$ 200 μg = 94·08	1·93
(6) 200	92·04		
o(7) 20	96·81	All 8 spots = 95·77	2·94
(8) 200	94·22		
Single spot (20 μg) on one sheet	97·70		

Actual radioactivities adjusted to dpm/50 μg for comparison purposes.

similar conditions of chromatography. Morrison and Orr (1966), Fairbairn and Suwal (1959) and Fairbairn and Wassel (1963) have reported that this is not always so, but in view of the unsuspected source of errors in preparing the initial spots, it was decided to re-investigate this problem using the machine designed to measure the initial volumes accurately. Ten sheets of paper each loaded with 10 spots of radioactive glucose ranging from 20 μg to 100 μg were run, either singly or in sets of four sheets simultaneously, in the same tank maintained at 18° in a constant temperature water bath. The details of the regression lines for each sheet are given in Table 4 and the two most divergent ones are shown in Fig. 2. A control series of initial spots containing the same quantities of glucose was prepared and the paper circles containing them transferred to phosphor for counting, without running them chromatographically. Their radioactivities were determined and the analysis of results is included in Table 4. Further details are given in the paper by Fairbairn and Relph (1968*b*).

Several conclusions can be drawn from these results. First, there seems no significant variation from tank load to tank load, even though some contained one sheet only and others contained four. Second, the correlation coefficients are all very near to unity indicating that for each sheet the measurements on the final spots are linearly related to the amounts of glucose contained in the initial spots. A third and important result is that the *slopes* do vary from sheet to sheet. The extreme values are shown in Fig. 2 and

TABLE 4

Details of regression lines relating radioactivities of the final spots (dpm) and quantities of glucose applied in the initial spots (μg). Ten sheets in four different tank loads were used. (Same solutions as in Table 3)

	Regression coefficient (Slope b)	Correlation coefficient (r)	Intercept $\times 10^{-2}$
1st tank load			
Sheet no. 1	200·9	0·9998	−1·80
2nd tank load			
Sheet no. 2	193·8	0·9910	2·73
Sheet no. 3	195·0	0·9998	−1·41
Sheet no. 4	193·5	0·9997	2·54
Sheet no. 5	198·7	0·9964	−1·03
3rd tank load			
Sheet no. 6	181·7	0·9997	5·00
Sheet no. 7	191·8	0·9996	3·06
Sheet no. 8	188·6	0·9990	−7·6
Sheet no. 9	185·1	0·9997	3·0
4th tank load			
Sheet no. 10	195·2	0·9986	−1·9
Control			
(initial spots without chromatography)	210·2	0·9997	1·69

The value for dpm at the midpoint of the regression line (60 mg) is about 117×10^{2} so that these values of the intercepts are only slightly above or below the origin.

statistical analysis of these regression lines shows that the difference in slope is significant. This therefore confirms earlier work by Fairbairn and Wassel (1963) and shows that, even within 10 sheets from the same batch of papers, significant differences in chromatographic behaviour occur. It is just possible that these variations between sheets are exaggerated by the radioactive counting technique used, but until alternative methods of investigation prove otherwise this source of error should be recognized. This can be done by using a standard on each sheet; at least two quantities (one a multiple of the other) should be used so that the slope can be checked. The unknown solutions should also be applied in pairs of spots differing from each other by the same factor.

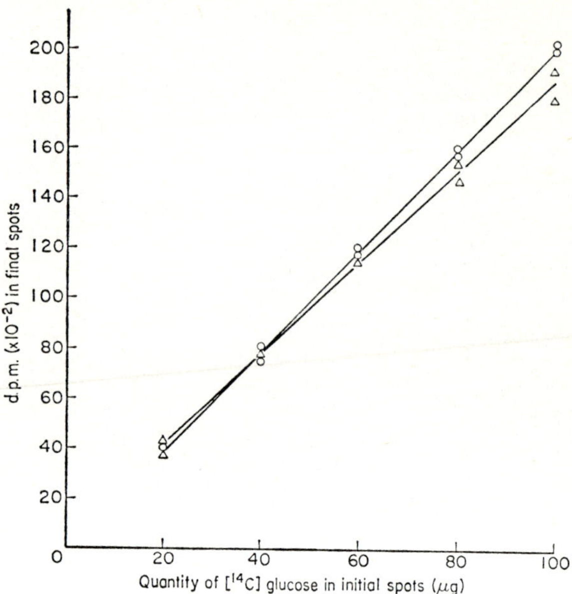

FIG. 2. Regression lines representing the extreme values of the slopes for the 10 sheets of paper referred to in Table 4 (i.e. sheet nos 1 and 6).

3. RETENTION ON THE STARTING LINE

Bush (1961) has shown that if the initial spots are dried vigorously some of the solute is irreversibly attached to the paper. Obviously this could lead to errors and to avoid this the initial spots were always dried in a current of cold air. In these circumstances there was no retention of glucose on the starting line. This may well apply to most substances but occasionally retention may occur. For example, 7–10% of T-morphine is retained during TLC on silica gel.

These experiments have, therefore, shown that provided the initial spots are of similar areas and are prepared from accurately delivered volumes of solution, the proportion of solute reaching the final spots will bear a constant relation to that of the initial ones. When differing quantities of substance occur in the initial spots the regression line, relating these quantities to those in the final spots, may vary in slope from sheet to sheet owing to differences in chromatographic behaviour arising from manufacturing variability. It is, therefore, necessary to apply both standard and test solution to each sheet for quantitative chromatography. Variation may be produced from tank to tank, though in practice this can be minimized by using the same tank, in identical conditions, for each tank load. In any event the use of standard and test solution on each sheet should significantly reduce errors from this source.

The final problem is the measurement of the quantity of substance in the final spots and this obviously introduces new potential sources of errors. However, it was decided to apply the information already obtained to the assay of certain alkaloids in plants using densitometric methods for evaluating the final spots, accepting the advice of Dr Franglen who has had a long experience in using the Chromoscan densitometer.

C. TREATMENT AND EVALUATION OF THE FINAL SPOTS

Densitometric methods were applied for this purpose to three groups of substance which are representative of probably quite a number which are more conveniently assayed by this means. The first group, the *Conium* alkaloids, are unsaturated compounds so that, even if successfully eluted, they could not be determined in small amounts by UVL spectrometry. A gas chromatographic method had been devised by Fairbairn *et al.* (1963) but it is not sufficiently sensitive. The second group, the opium alkaloids, can be assayed by paper chromatography involving area measurements (Fairbairn and Wassel, 1963), but though a reliable method it is rather tedious. In the third group, the *Mitragyna* alkaloids, a TLC system has been devised by Shellard and Alam (1968) which gives reddish spots against a white background and therefore is very suitable for densitometry. This also enabled us to apply our principles to TLC.

1. QUANTITATIVE PAPER CHROMATOGRAPHY

A summary of the chromatographic conditions necessary to produce the uniform final spots already described, and to achieve minimum error in measuring their content by densitometry, are as follows.

a. Preparation of the Papers

i. Paper. Kept free from dust in sealed (and re-sealed) packets; allowed to equilibrate in the tank overnight.

ii. Initial spots. Delivered by the special machine and dried by a current of cold air, in agreement with the opinions of Bush (1963) and Consden (1964).

b. Chromatographic Run

The tank was immersed in a large, constant temperature water bath and a critical solvent volume of 1 % used throughout (Block *et al.*, 1958). Descending technique was used and the initial spots placed about 11 cm from the upper edge of the paper. This allowed ample space for the edge to be dipped in the trough and the mobile phase to have achieved even flow rate before it reached the spots. The solvent front was then allowed to reach a constant distance from the starting line during the run; this was conveniently arranged by punching one or two holes at the prearranged distance and observing when these were reached by the solvent front. The paper was removed and rapidly dried by a current of warm air so as to "fix" the final spots as quickly as possible.

i. Colour development. It was found that spraying on both sides of the paper is better than dipping largely because, when the latter was used with our particular reagent (platinum potassium iodide), surface scum attached itself to the paper producing black speckles. Obviously with other reagents dipping may be more convenient (Bush, 1963; Franglen, *cf.* p. 17). The sprayed spots were dried in a stream of cold air.

c. Densitometry

One of the main problems here is that of background colour due to excess of reagent on the paper not containing the spot. Although the use of selective light filters helps considerably the background will almost certainly produce some reading in the instrument and this value will vary with amount present and colour changes due to air oxidation and fading. To avoid this densitometry was commenced exactly one hour after colour development. If this was not convenient, the strips containing the final spots were stored without colour development between folds of chromatographic paper in the dark and sprayed when convenient.

2. RESULTS

a. Opium Alkaloids

A series of eight initial spots was produced on paper by the methods already described. Each spot was produced from 2·4 μl. of a solution containing morphine (8·1 μg), codeine (4·1 μg), and thebaine (3·2 μg). Four such sheets were prepared, run chromatographically and treated as already described. The results are shown in Table 5.

TABLE 5

Densitometry of some opium alkaloids; average peak height (in brackets) and coefficient of variation for eight spots on each sheet of paper

	Morphine	Codeine	Thebaine
Sheet no. 1	(87·3 mm) ± 1·9%	(27·0 mm) ± 3·21%	(46·5 mm) ± 2·63%
Sheet no. 2	(85·8 mm) ± 2·13%	(27·8 mm) ± 4·06%	(45·1 mm) ± 2·37%
Sheet no. 3	(85·1 mm) ± 1·58%	(27·4 mm) ± 3·19%	(51·0 mm) ± 1·90%
Sheet no. 4	(81·0 mm) ± 2·14%		

It can be seen that the variations found are similar to the earlier ones based on radioactive counts: the results for codeine are less consistent and this may be due to additional errors in measuring low peak heights.

b. Conium Alkaloids

The two major alkaloids of *Conium maculatum* (the Poison Hemlock) are coniine (2-*n*-propyl-piperidine) and γ-coniceine (2-*n*-propyl-Δ'piperideine). Their R_F values according to Cromwell (1956) are 0·66 and 0·32 respectively. Sheets containing a series of initial spots of increasing quantity were chromatographed and treated as already described. The results are given in Table 6 and again confirm earlier ones based on radioactive counts in that the coefficients of variation for replicates are low, the correlation coefficients of the regression lines are satisfactory but the slopes vary from sheet to sheet.

TABLE 6

Densitometry of some Conium alkaloids

Replicates			
Coniine	(7 × 4·8 μg)	(65·7 mm)*	± 1·19†
γ-Coniceine	(7 × 8·1 μg)	(69·4 mm)*	± 1·90†
Regression lines			
Coniine	(6–12 μg)	$b = 8\cdot7750$	$r = 0\cdot9939$
γ-Coniceine	(3·4–13·4 μg)		
	Sheet no. 1	$b = 1\cdot1952$	$r = 0\cdot9908$
	Sheet no. 2	$b = 0\cdot8900$	$r = 0\cdot9966$

* Average peak heights.
† Coefficients of variation.
b = Slope; r = correlation coefficient.

3. THIN-LAYER CHROMATOGRAPHY

A series of five spots formed from 6, 8, 10, 12 and 14 μl. of a solution of isomitraphylline (0·76 μg/μl.) was delivered on to each of two plates which were treated chromatographically by the method of Shellard and Alam (1968). Two further plates were prepared, each containing a series of five spots formed from 16, 18, 22, 24 and 26 μl. of the solution. Densitometry was carried out with a Chromoscan having a thin layer attachment and the results are given in Table 7. Once more these results confirm our earlier work based mainly on paper chromatography.

TABLE 7

Densitometry of Isomitraphylline using TLC

	Plate 1	Plate 2		Plate 3	Plate 4
6 μl.	16 mm	13·5 mm	16 μl.	32·0 mm	32·5 mm
8 μl.	20 mm	19·5 mm	18 μl.	34·0 mm	34·0 mm
10 μl.	24 mm	24·0 mm	22 μl.	36·0 mm	36·0 mm
12 μl.	29 mm	27·5 mm	24 μl.	37·0 mm	37·0 mm
14 μl.	31 mm	29·5 mm	26 μl.	39·5 mm	40·0 mm
b	2·57	2·63		1·184	1·184
r	0·9938	0·9823		0·9938	0·9878

Ethanol solution 0·76 μg/μl. used. Peak heights in mm.
b = Slope; r = correlation coefficient.

CONCLUSION

The most important result of this work has been to demonstrate that one major source of error is the measurement of the volumes of solution to form the initial spots. If this is done accurately the following other sources of error can be satisfactorily dealt with. The initial spots should be of equal area and should be dried in a current of cold air. The tank conditions should be as constant as possible (see text). During the chromatographic run no significant error occurs due to lateral diffusion provided that adjacent spots are not too dissimilar in solute content. As the slopes of the regression lines relating quantities in the final spots to those in the initial ones vary from sheet to sheet (and almost certainly from thin-layer to thin-layer), standard spots should be used for each sheet. Two quantities of the standard, one a multiple of the other, and two quantities of the unknown, differing by the same factor, should be used for each assay. When these principles are used it is possible to reduce the error of individual determinations to about $\pm 6\%$ (coefficient of variation about 2–3%). As pointed out at the beginning of the chapter, all the ranges of error have been based on individual results; obviously in practice most workers in quantitative chromatography use means of replicates. In these circumstances the limits of error will be reduced by $1/\sqrt{n}$, where n is the number of replicates made. With the machine already referred to, the four spots necessary for each assay can be delivered automatically, with three to four sets on each sheet. Thus on one sheet of paper enough data can be obtained to produce an accurate result.

It is probably unnecessary to point out that the results reported here are based on only a few chromatographic systems and further experience is necessary before they can be claimed to be of general applicability. Although preliminary work with densitometry is encouraging, clearly a new set of variables is introduced when such attempts to estimate the quantities in the final spots are made.

ACKNOWLEDGEMENT

I would like to thank Mr S. J. Relph for carrying out much of the practical work described in this chapter (see Relph (1968)).

REFERENCES

Block, R. J., Durrum, E. L. and Zweig, G. (1958). "A Manual of Paper Chromatography and Paper Electrophoresis", 2nd edn., p. 61. Academic Press, New York.

Bush, I. E. (1961). "Chromatography of the Steroids." Pergamon Press, London.

Bush, I. E. (1963). *Meth. Biochem. Anal.* **9**, 149.

Bridger, V. E. and Relph, S. J. (1967). *British Patent Appl.* Nos 54215/65 and 40136/66.

Consden, R. (1964). *2nd Intern. Chromatog. Symp.*, p. 164. Elsevier, Amsterdam.

Cromwell, B. T. (1956). *Biochem. J.* **64**, 259.

Fairbairn, J. W. and Relph, S. J. (1968a). *J. Chromatogr.* **33**, 494.

Fairbairn, J. W. and Relph, S. J. (1968b). *J. Chromatogr.* In press.

Fairbairn, J. W. and Suwal, P. N. (1959). *Pharm. Act. Helv.* **34**, 561.

Fairbairn, J. W. and Wassel, G. (1963). *J. Pharm. Pharmac.* **15**, 216T.

Fairbairn, J. W., Shellard, E. J. and Talalaj, S. (1963). *Planta Medica* **11**, 92.

Fisher, R. B., Parsons, D. S. and Holmes, R. (1949). *Nature, Lond.* **164**, 183.

Genest, K. and Farmilo, C. G. (1959). *J. Am. pharm. Ass.* **48**, 286.

McEvoy-Bowe, E. and Lugg, J. W. H. (1961). *Biochem. J.* **80**, 616.

Ohtsu, T. and Mizuno, D. (1952). *Jap. J. med. Sci. Biol.* **5**, 37.

Morrison, J. C. and Orr, J. M. (1966). *J. pharm. Sci.* **55**, 936.

Pataki, G. (1967). *Chromatogr. Rev.* **9**, 23.

Relph, S. J. (1968). "Studies in Quantitative Layer Chromatography." Thesis, University of London.

Römisch, H. (1961). *Pharmazie* **16**, 373.

Shellard, E. J. and Alam, M. Z. (1968). *J. Chromatogr.* **33**, 347.

Chapter 2

Quantitation of Paper Chromatograms

GEOFFREY FRANGLEN

*Department of Chemical Pathology, St. George's Hospital Medical School,
University of London, London, England*

The quantitation of paper chromatograms forms an extremely complex problem, and the successful methods used in its solution vary markedly with the nature of the material under analysis. Nevertheless, there are general principles which can be usefully applied. Particular reference will be paid in this chapter to the classical type of paper chromatogram in which the originally colourless compound is revealed after its chromatographic separation by treatment with a reagent which converts it to a coloured, visible derivative. Quantitation is then effected through estimation of this derivative.

The problems of quantitation can be divided into three parts:

    (1) the formation of the coloured derivative;

    (2) its estimation;

and (3) any special conditions relating to the chromatographic separation.

The conditions under which the coloured derivative is formed are far from ideal, since the compound is spread on paper fibres into which it penetrates to a more or less limited extent. The reagent is coated onto it by dipping or spraying the chromatogram, and there is no guarantee that this will penetrate completely either into the paper or into the dried compound. Moreover, the reaction often takes place in a dry or nearly dry state. Even when the colour has been formed successfully, the derivatives too often are unstable and fugitive, or the reactions which lead to their formation may vary considerably with the concentration of the original compound on the chromatogram. In the circumstances it is almost surprising that the quantitation of paper chromatograms can be as good as it frequently turns out to be. Obviously, accuracy will vary considerably with the class of compound under analysis and with the particular techniques used, but, to give two examples, it is comparatively easy to estimate the common amino acids from a protein hydrolysate by one-dimensional chromatography with an accuracy of about 1–2%, though by using special techniques some workers have claimed down to about 0·2%. Sugars can be estimated quite easily and quickly on a paper chromatogram with an accuracy of about 3%.

In qualitative paper chromatography it is common to use comparatively low concentrations of detecting reagent. This is done deliberately to reduce the background which may be due to the colour of the reagent itself or may gradually develop with time. In this way detection of separated components is facilitated by maintaining a high visual contrast between their colour and the surrounding blank paper. In such circumstances the weak spots of the class under analysis react reasonably fully to form the coloured derivatives, but the stronger ones are only partially converted (Fig. 1). In quantitative

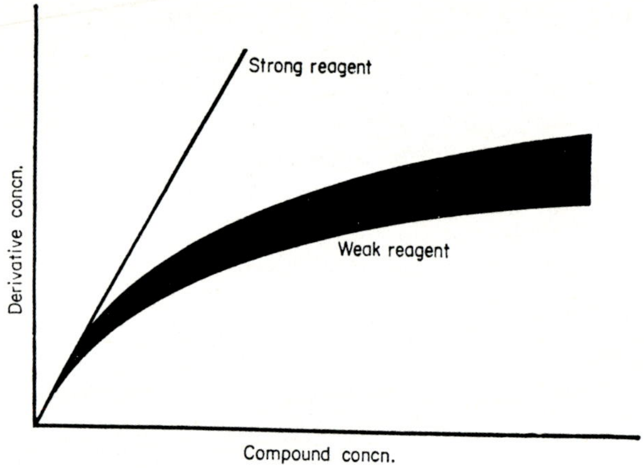

FIG. 1. Effect of strong and weak reagents on the production of coloured derivative from a compound on a paper chromatogram.

terms this results, of course, in a decreased sensitivity at higher concentrations of the compound, but, much more important, it usually leads to a considerable variation in the amount of derivative formed at these higher levels. It is essential, therefore, that the concentration of reagent should be sufficient to deal adequately with the greatest concentration of compound which is likely to occur. When this is checked for any particular method, what is usually found is that the amount of coloured derivative formed increases more or less linearly with the concentration of the reagent until a constant level is attained (Fig. 2). The concentration selected for quantitative work should lie well within the range where this occurs. The differences between the qualitative and quantitative versions of a reagent can be quite marked, for example, amino acids are usually detected on paper chromatograms with a 0·1–0·2% solution of ninhydrin, but for quantitation a solution of at least 1% must be used, an increase of 5–10 times.

Unfortunately, when the concentration of the reagent is increased, the colour of the background is usually increased as well, and the problem of keeping it even becomes acute. In all forms of quantitative paper chromatography, whether using elution or densitometric techniques, the empty areas

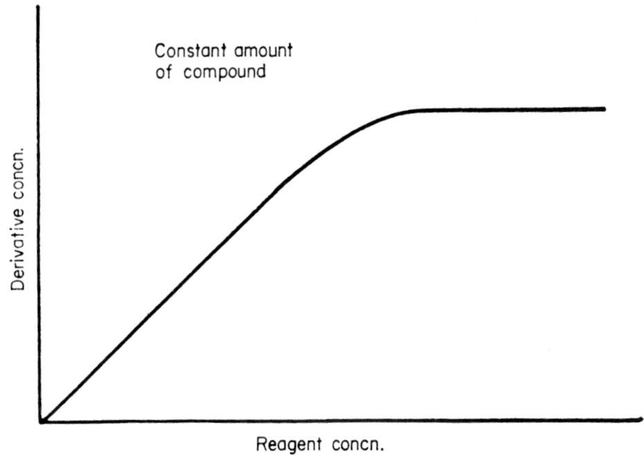

FIG. 2. Effect of reagent concentration on the production of coloured derivative from a constant amount of compound on a paper chromatogram.

of the chromatogram form the reagent "blank", and it is essential that this should be as constant as possible. The reagent is normally applied in one of two ways, either by spraying or by dipping the chromatogram through it. It is essential with spraying to provide the necessary pressure either from an air-pump or from a gas cylinder; mouth pressure or a hand-operated rubber bulb do not provide sufficiently even coverage. If some form of illumination can be safely provided behind the chromatogram, this is a considerable help towards even spraying. Dipping the chromatogram through the reagent usually gives superior results, especially if the reagent can be made up in some readily volatile solvent such as acetone or ethanol. If there is a choice, the solvent used for the reagent should be one in which the compound under analysis is not readily soluble. When aqueous reagents are used it is usually necessary to blot any excess off the dipped chromatogram, and this generally needs a certain amount of practice to get an even distribution of reagent over the whole area. It is better in such cases to pass the wet chromatogram through a small wringer kept for the purpose, though this obvious solution can bring its own problems.

The speed of reaction between reagent and compound varies considerably from case to case. When it is rapid and largely completed before the chromatogram dries, a high proportion of derivative and a low level of background colour is usually the result. Slow reactions can be speeded in some cases by heating the chromatogram, but, whilst this is useful quite often in qualitative

     G. FRANGLEN

work, it must be used with care in quantitation. It often results in a marked increase in background colour, and generally it is better to allow the colour to develop at room temperature if it will do so. Many derivatives are sensitive to some extent to light and it is usual, therefore, to keep the chromatogram in the dark during the reaction period. What is absolutely essential is some form of protection against the vagaries of the laboratory atmosphere. An efficient method is to store the chromatogram in an envelope cut from 0·015 in. polythene sheeting with the edges stapled together and sealed with ordinary drafting tape.

When the reaction is slow, the period allowed for it has to be timed. There are two reasons for this:

(1) The background often starts to develop or intensify its colour after a certain period. For maximum sensitivity the time of reaction has to be such that the maximum difference between the colour of the spot and of the background is obtained (Fig. 3).

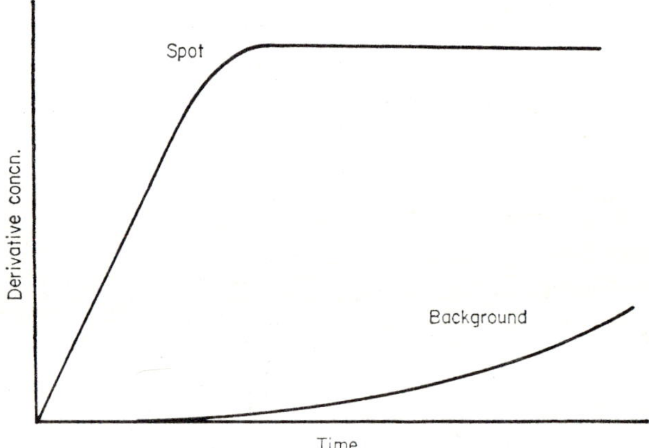

FIG. 3. Comparison of the effect of time on the production of coloured derivative from a spot of compound and from the background of a paper chromatogram.

(2) The spot itself frequently starts to fade after attaining its maximum colour and thus for maximum sensitivity estimation of the derivative must be made during the period of maximum colour formation (Fig. 4). In some cases the derivative can be preserved or converted to a more stable form by spraying a second reagent onto the developed chromatogram or by including suitable compounds in the original reagent; amino acids for example, form a purple colour with ninhydrin on paper, and such spots generally begin to fade after about a day or so at room temperature. They may be preserved by spraying the chromatogram with nickel acetate, or, better still, if cadmium acetate is included in the original ninhydrin reagent,

the purple colour as it is formed is immediately converted to a crimson derivative which is stable on the chromatogram for at least five years.

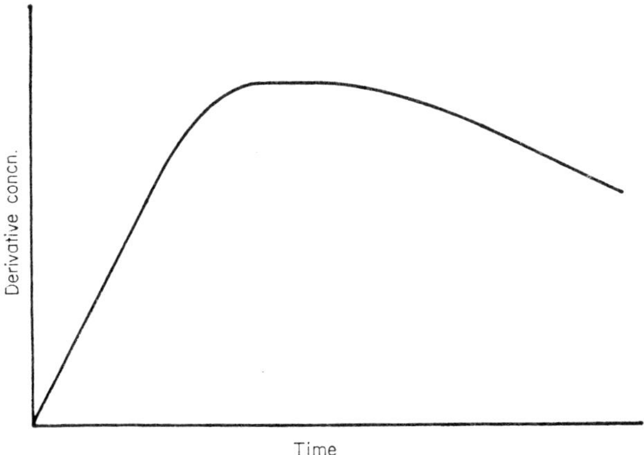

Fig. 4. Effect of time on the production of coloured derivative from a spot of compound on a paper chromatogram.

When the colour is particularly fugitive, the best means of quantitation is through a photograph of the chromatogram taken at the time of the maximum intensity of the spots. Suitable filters should be used on the camera to improve contrast, and quantitation can be made either through area measurement of the spots or by densitometry. If the latter technique is used, it is essential to have negatives and prints of fairly low contrast—in photographic terms, to keep off the toe and shoulder of the density-exposure curve and to stay strictly within its straight-line portion.

During the formation of the coloured derivative one of three types of interaction may occur (Fig. 5). It may be that the colour produced is in exact proportion to the amount of material applied, suggesting that the compound has reacted completely. This is usually restricted to low concentrations of compound and is thus limited in its application. Obviously it tends to occur more often with two-dimensional analyses where the effects leading to dilution of the original material are greater than in the cases of one-dimensional systems. More often the production of the derivative may appear to be more or less linear at low levels of compound, but deviates more and more as the concentration rises. In most cases this is simply a matter of the reagent being unable to penetrate sufficiently into the layers of dry material deposited on the paper fibres. This effect can sometimes be reduced by changing the solvent of the reagent to effect greater penetration, for example, by using chloroform instead of acetone, thus reducing the rate of evaporation of the solvent from the paper and perhaps thus giving the reagent the necessary time to soak into the layers of compound. Hanging the chromatogram in a damp atmosphere can also help. The important point to note, however, is that, whilst such reactions are not complete at higher concentrations, they are often extremely reproducible when care is taken. A linear reaction is convenient, but in this type of work it is not essential for accuracy.

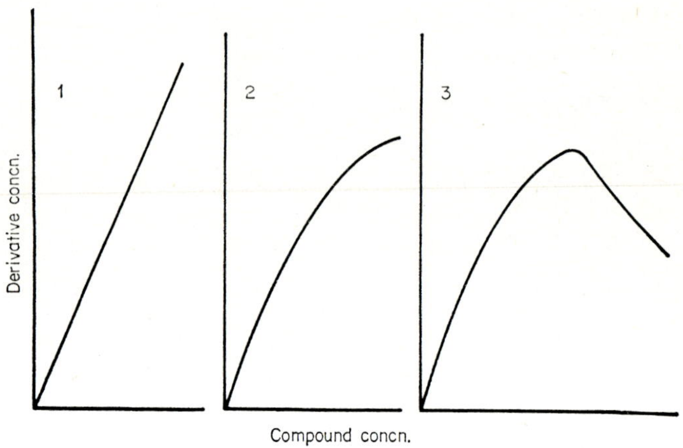

FIG. 5. Three common types of reaction curves between compound and coloured derivative concentrations on paper chromatograms.

A third type of reaction between compound and reagent is anomalous, but unfortunately is one which seems to occur rather often in drug chromatography. In this type low levels of material react normally or perhaps demonstrate the insufficient penetration of the reagent into the material similar to the second type. Above a certain concentration, however, a completely new phenomenon occurs. Not only does the reagent not appear to penetrate the material, but the concentration of the latter is sufficient to modify or even inhibit the reaction, so that there is an actual decrease in the amount of colour formed as the concentration is increased further. This effect can be seen on a chromatogram when the low concentration spots and the edges only of the high concentration spots react to give the normal colour of the derivative. The centres of the high concentration spots either react minimally, so that the spot has more the appearance of a ring, or else have a distinctly different colour to that of their edges. Obviously, this behaviour considerably restricts the methods that can be used for the quantitation of such chromatograms.

There are basically three methods for the quantitation of paper chromatograms (Block *et al.*, 1955):

(1) The derivative can be eluted from the paper and its concentration estimated in solution by absorptiometry.

(2) The area of the spot can be measured.

(3) Direct densitometry of the spot on the paper can be effected.

A fourth type should also be mentioned, that of making comparisons of standard and test spots by eye. It is a somewhat rough method, but under optimum conditions can attain an accuracy of perhaps $\pm 25\%$. It is really a combination of spot area assessment combined with a sort of ocular densitometry.

These three main methods, elution, spot area measurement and densitometry will give good accuracy if and only if certain conditions are rigorously observed:

(1) The test solution must be analysed together with appropriate standards. In one-dimensional analyses these should all be included on the same sheet; in two-dimensional analyses they should all be included in the same batch of chromatograms. The size of the spots applied to the chromatogram, both for test and for standards, must be the same throughout; this is essential for accuracy. During chromatography the spots diffuse and their contents become diluted and as already mentioned this will affect the amount of derivative subsequently formed. Thus it becomes essential to standardize this diffusion and dilution. The apparatus designed by Bridger and Relph obviously has enormous potential in providing starting spots of regular size. Difficulty can arise when, as is so often the case, the test solution is weak and needs several applications to the chromatogram to attain a suitable load. The over-enthusiastic use of a hair-dryer can desiccate the paper sufficiently so that subsequent applications either to the original spot or elsewhere on the start-line are taken up more rapidly and form smaller circles.

(2) The concentrations of the standards applied to the chromatogram must straddle the concentration of the test solution. There must be no extrapolation of the line formed by the standards in order to determine the concentration of the test. This should not be done even in those cases where it seems that the reaction between compound and reagent is complete, since the method of measuring the derivative may impose a curvilinear relationship on the system.

(3) The treatment of test and standards must be identical throughout the analysis. This would seem obvious, but, unfortunately only too often descriptions of methods are reported in which this is not done. The standard solutions are run separately or differently, or perhaps are even retained for future analyses, and in some cases are used directly from the stock bottle without the necessary intervening step of chromatography as a check on losses due to adsorption to the paper, fading of the derivative or to other factors.

Elution of the derivative can form one of the most accurate methods of quantitation, and levels of accuracy of below 1% are frequently claimed.

Obviously the test material must be completely separated from other spots on the chromatogram, and for this reason this type of analysis is more common with two-dimensional systems. The prime disadvantage of elution is that, comparatively speaking, rather large quantities of material are usually needed to obtain a solution of sufficient strength to measure accurately. Even when using microcells in the absorptiometer it may be necessary to apply the material as long lines rather than as spots, or to overload the chromatogram. Various minor points should be noted, (i) ideally the colour should be eluted completely, but this is sometimes difficult since paper acts as a weak adsorbent as well as providing a partition system for chromatography, (ii) some form of agitation is usually needed during elution and this often fragments the paper into fibres which have to be removed before measurement of the solution, (iii) quite often solutions of the derivatives are surprisingly unstable and have to be determined rapidly and (iv) the chromatographic solvent can affect the development of the background blank in spite of the care taken to apply the reagent evenly. It may be found that the background varies over the total area of the chromatogram and in such cases blanks must be taken from clear areas as close as possible to the position of the test spot. A particularly useful aspect of elution techniques is that quite often additional reagent may be included in the eluting fluid, so that the compound is completely converted to the derivative in spite of an incomplete or anomalous reaction on the chromatogram.

Quantitation by measurement of spot area provides moderate accuracy; methods using this technique usually attain levels of about 5–10%. When standards of initially the same small size are compared, the area of the spots after chromatography is closely proportional to the logarithm of the compound concentration over a fairly short range of concentrations (Fig. 6). The area can be determined by planimetry but it is usually sufficient to measure the major and minor axes of the spot and to take their product as a measurement of area. With diffuse spots it is often difficult to decide exactly where the edge starts. In such cases it is useful to make a copy of the chromatogram onto high contrast photographic paper such as that used for document copying.

Densitometry is one of the most convenient and sensitive methods of quantitation and it is comparatively easy to measure nanomoles of material with an accuracy of 1–2%. Briefly it is a direct determination of the concentration of the derivative on the chromatogram by a form of absorptiometry. The chromatogram is scanned with a beam or spot of filtered light and the proportion of this either reflected or transmitted by the paper is measured photoelectrically (Ayers, 1956; Crook, 1956). In a way, therefore, the approach is similar to that of conventional absorptiometry, and it might be thought that a relationship similar to that of Beer's Law might apply.

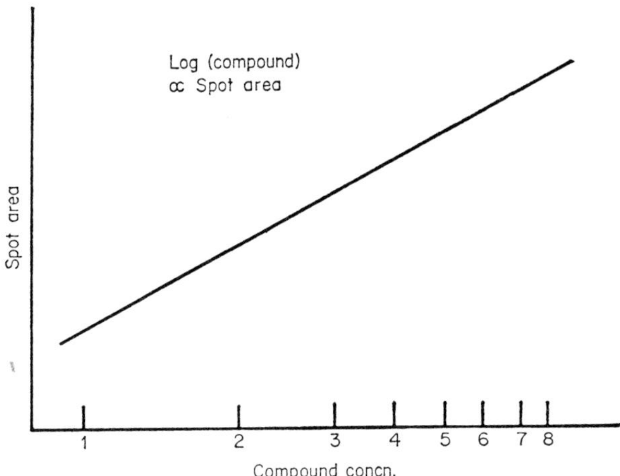

FIG. 6. Relationship between the area of a spot on a paper chromatogram and compound concentration.

Certainly the instrument manufacturers appeared to think so initially, since the majority of the commercially available instruments obey Beer's Law rather well in their basic form, i.e. without the insertion of correction cams. This is satisfactory when the supporting medium is completely transparent, but paper, at best, is only translucent. The relationship between optical density and colour concentration on a solid medium was examined by Kubelka (1948) and Kubelka and Munk (1931). Kubelka derived two expressions to describe it, one for the case where the light falls onto the dyed surface and is then measured after reflection, and the other when it is measured after transmittance through a perfectly diffusing thickness of the medium. In the formula for reflectance it is assumed that the supporting medium is infinitely thick, or at least so thick as to make no essential difference. In such circumstances R_∞, the reflectance, can be related simply to the coefficient of absorption (K) and the coefficient of scatter (S). Within certain limits the colour and the medium should affect the light independently of each other, and so the ratio K/S should be proportional to the dye present. The dye concentration (c) in a given volume of medium should be proportional to the total K/S ratio minus the corresponding blank K/S. In transmission the thickness of the material must be taken into account, and so the expression, although obviously related to the previous one, is somewhat more complicated. Both expressions give a relationship between optical density and dye concentration which is distinctly different to that derived from Beer's Law (see appendix p. 28). Within limits they overlap the Beer's Law relationship at low concentrations of dye, but as this increases, the resultant optical density falls off rapidly and the curve resembles more a hyperbola than the straight line given by Beer's Law.

It might be thought, therefore, that Kubelka's Laws should be incorporated in these densitometers rather than having them based on Beer's Law. In practice, however, such an approach would not give a linear response to derivative concentration since, in Kubelka's terms, paper is not an ideal medium. It is not "infinitely" thick nor perfectly diffuse; under microscopic examination it appears as an uneven, intensely scattering medium, full of spaces and fibres, fibres which are partially crystalline, partially amorphous. The distribution of colour even in a strong spot is very patchy. In the circumstances, therefore, the incorporation of a Beer's Law response into the ordinary densitometer is as good as anything else, and, in fact, usually gives convenient relationships between the instrumental response and the colour concentration, even if they are not linear.

With both reflectance and transmittance densitometers it is necessary to oil the paper before scanning, partly to cut down the light scatter and partly to ensure that all the thickness of the derivative plays its part in modifying the light. Special mixtures, usually containing methyl salicylate, have been described with refractive indices close to that of the paper fibres, but in practice ordinary liquid paraffin is perfectly satisfactory, and is to be preferred.

The densitometer can be used in one of two ways for quantitation. In the first method the spot can be scanned to give a curve representing the variation of colour in it over the path of scanning, and this curve can then be integrated to give a measurement of the amount of derivative present in the total spot. This method imposes certain limitations and the spot must be clearly separated from any others in order to avoid unwanted contributions to the result. The spot must either be scanned completely by repeated passages through the instrument, or it must always be scanned in exactly the same place each time, say, through the centre along the line of solvent flow. Unfortunately spots do not always line themselves up conveniently, even in careful one-dimensional chromatography. Some workers have tried to simplify this problem by restricting lateral diffusion of spots by running the material as bands either on narrow strips or on chromatographic "doilies". For quantitative work this is unnecessary since the other method of densitometry is so much better.

In the second method of using the densitometer, the spot is scanned repeatedly, the path of the scanning beam being slightly shifted each time (Franglen, 1963) until the maximum density of the spot is found, which is indicated by the maximum height attained by the curve above the base line. This "peak-height" is proportional to the logarithm of compound concentration, a similar relationship to that for spot area (Fig. 7). It is, however, much more accurate for quantitation and can be used to estimate considerably smaller amounts of materials. It is not affected by slight overlapping of the spots, i.e. when this does not affect the maximum density at the centre of the spot. It is best to use this method with one-dimensional systems, when standards and test line up next to one another and can be excised from the chromatogram as a single strip for insertion in the densitometer.

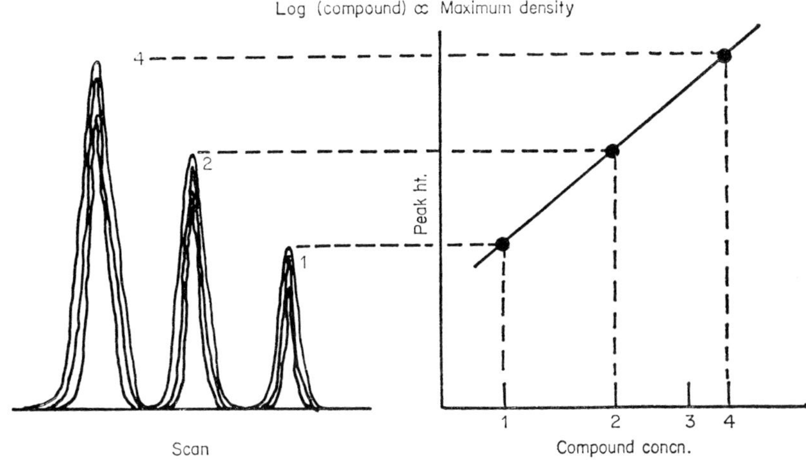

FIG. 7. Relationship between the maximum density of a spot on a paper chromatogram and compound concentration.

Summarizing, the quantitation of paper chromatograms is possible to a high degree of accuracy, provided that certain precautions are taken. The chromatograms must contain both test material and suitable standards and the treatment of these must be strictly identical throughout the analysis. The formation of the coloured derivative should be maximal and tightly reproducible. Once formed, it may be estimated by the techniques of elution, spot area measurement or densitometry. Of these, elution techniques usually have the highest accuracy but need comparatively large amounts of material. Densitometric methods, although slightly less accurate, have the advantage of needing much smaller amounts of material for analysis, and, in most cases, should be considered the method of choice.

REFERENCES

Ayers, C. W. (1956). *Mikrochim. Acta* **9**, 1333.
Block, R. J., Durrum, E. L. and Zweig, G. (1955). *In* "A Manual of Paper Chromatography and Paper Electrophoresis". Academic Press, New York.
Crook, E. M. (1956). *In* "Ciba Foundation Symposium on Paper Electrophoresis", p. 132 (Wolstenholme, G. E. W. and Millar, E. C. P., eds.). Churchill, London.
Franglen, G. (1963). *Joyce-Loebl Rev.* **1**, 6.
Kubelka, P. (1948). *J. opt. Soc. Am.* **38**, 448.
Kubelka, P. and Munk, F. (1931). *Z. techn. Physik*, **12**, 593.

APPENDIX

KUBELKA'S LAWS FOR DENSITOMETRIC MEASUREMENTS OF
COLOUR ON SOLID MEDIA

Reflectance

$$R_\infty = a - b$$

where R_∞ = monochromatic reflectance of a material of such thickness that a further increase in thickness does not alter the reflectance reading

K = absorption coefficient

S = scattering coefficient

$a = 1 + K/S$

$b = (a^2 - 1)^{\frac{1}{2}}$

Within limits the colour and medium absorb and scatter light independently. The above expression gives the ratio K/S:

$$K/S = \frac{(1 - R_\infty)^2}{2R_\infty}.$$

Since K/S for the colour is proportional to the amount of colour present, it follows

$$k.c = \left[\frac{(1 - R_\infty)^2}{2R_\infty}\right]_{\text{total}} - \left[\frac{(1 - R_\infty)^2}{2R_\infty}\right]_{\text{medium blank}}$$

where c = colour concentration

k = a constant

Transmission

$$T = \frac{b}{a \sinh bSx + b \cosh bSx}$$

where x = thickness of medium

T = transmission

Chapter 3

Quantitative Thin-Layer Chromatography using Elution Techniques

WILLIAM E. COURT

*School of Pharmacy, College of Technology,
Liverpool, England*

In the field of pharmaceutical analysis a common problem is the assessment of complex mixtures occurring in natural products and pharmaceutical compositions. The introduction of thin-layer chromatography provided a quicker and more elegant technique for the separation of chemical components than was provided by the earlier chemical methods. Inevitably this new technique was soon widely exploited and adapted for quantitative analysis.

The separation of the chemical constituents by thin-layer chromatography, recovery by quantitative elution, and subsequent measurement with a suitable instrument would appear simple but, in practice, this procedure involves a considerable number of variable factors. Randerath (1963) stated that the most accurate results in quantitative thin-layer chromatography were obtained by removal of the compounds to be analysed from the adsorbent using suitable solvents (elution) followed by spectrophotometric or colorimetric estimation. In this contribution the variable factors, the overall accuracy, the advantages and the disadvantages of the method are considered and some reported applications reviewed.

The elution method can be considered under the headings:

(A) The separation system.
(B) The application of the test and reference samples.
(C) The recovery of the assay and reference samples.
(D) Measurement methods.
(E) Accuracy and applications.

A. THE SEPARATION SYSTEM

The initial requirement of the elution technique is a separation system, which will facilitate the division of the unknown mixture into clearly defined areas which can be removed quantitatively from the plates, eluted and estimated. Thin-layer chromatography is essentially a procedure in which a solution of substances to be separated is passed over a finely-divided and activated solid in a specified direction for a specified distance. As the components in the liquid carrier phase are usually retained differentially on the solid stationary phase, separation occurs. For quantitative work this separation must be both adequate and readily reproducible.

It has been shown that the R_f value of a substance depends on the following factors (Brenner *et al.*, 1962; Shellard, 1964; Geiss *et al.*, 1965):

1. Adsorbent. (*a*) Quality and nature (impurities, batch variations).
                (*b*) Uniformity of the layer.
                (*c*) Thickness of the layer.
                (*d*) Activation and storage of the plates.
2. Solvent.   (*a*) Quality and nature (impurities, stability).
                (*b*) Degree of saturation of the solvent vapour.
                (*c*) Temperature.
3. Solutes.   (*a*) Amount of solutes, i.e. load.
                (*b*) Nature of the solutes (stability).
4. Technique. (*a*) Size of chambers or tanks.
                (*b*) Development distance, i.e. distance from the starting line to the final solvent front.
                (*c*) Direction of development.

1. ADSORBENT

Normally the factors concerning the adsorbent are considered with respect to R_f values and suitable separations of substances. In elution techniques it is also necessary to anticipate problems arising from the adsorbent interfering with subsequent stages in the assay procedure. It must be practicable to recover the required solute readily and quantitatively from the adsorbent. Adsorbent layers employed must be uniform and firm as losses may be incurred during manipulation of loaded plates if the surface is loose and powdery. Particle size, a factor of importance for uniform development of plates, may provide difficulties in elution if the particle size is too small. Thus, in our laboratories, adequate separations of alkaloids have been effected on plates prepared with magnesium trisilicate B.P. as adsorbent, but the finer particles could not be satisfactorily removed from the resultant eluates even after prolonged centrifugation and, therefore, such eluates were unsuitable for spectrophotometric estimation.

Binding agents are normally necessary to yield firm, uniform layers, but, in elution methods, such agents are a potential cause of inaccuracy due to their solubility in the eluting solvent. Honegger (1962) demonstrated that a significant extraction residue was obtained from chloroform, acetone and benzene. Silicagel free of gypsum also yields an inorganic extraction residue, especially if silicagel of particle size smaller than 40 μ is present as a binder.

Centrifugation may remove such particulate matter, or column filtration or change of solvent may be adopted (Bobbitt, 1963) although such procedures introduce further potential manipulative losses. To minimize interfering substances, Brown and Benjamin (1964) advocated washing silicagel layers with 20% v/v ethanol and 80% v/v methanol at right angles to the direction of flow and drying in an oven at 110°C before use. Prewashing has also been recommended by other workers (Kirchner *et al.*, 1954; Stanley *et al.*, 1957; Stanley, 1959).

Reproducibility of separation can only be attained if the nature of the adsorbent is constant. Thus extraneous mineral matter, e.g. traces of iron, must be minimal, especially if the final process is a colorimetric estimation. Teichert *et al.* (1961) noted that iron traces interfered with the colour complex produced by morphine. Iron traces can be removed by treating the adsorbent with boiling ethanol containing some sulphuric acid or by a preliminary development of the thin-layer plates with methanol: concentrated hydrochloric acid (9 : 1) followed by drying and reactivation of the layers (Merck, 1962).

Standardized procedures of activation and reactivation are prerequisites for effective reproducibility. Kelemen and Pataki (1963) advocated leaving Kieselgel G (silicagel) plates overnight at an ambient temperature rather than oven-drying at 110–120°C for 30 min. prior to cooling in a desiccator. Although Pataki and Keller (1963) stated that R_f values on Kieselgel G layers increased with layer thickness in the range 0·25–1·0 mm, Dallas (1965) considered layer thickness and the angle of slope of the plates in the tank as the least important factors, noting also that the temperature and the distance of solvent travel were not very important; the adsorbent activity and the method of development were extremely important. As adsorbent activity is affected markedly by water present, Dallas recommended equilibration of thin-layer plates over salt solutions to yield standard conditions on the plates.

It is important, therefore, to adopt standardized methods of preparing and handling the thin-layer plates and, as far as possible, to use adsorbent from the same batch. Experience obtained in separating *Rauwolfia* alkaloids qualitatively and quantitatively on silicagel G layers suggests that prewashing is less important than a rigidly observed routine for the preparation and handling of plates.

2. SOLVENT SYSTEM

The solvent system employed must obviously afford adequate separations of the components of the mixture to be assayed, but careful choice is necessary if the developed and dried plates are subsequently to be eluted. Efficient removal of the solvent system during drying ensures the absence of traces of solvent which may interfere with subsequent spectrophotometric determinations. Plates, highly activated after drying, should be handled carefully as readsorption of solvent vapour is possible. Harris (1966) observed that silicagel thin-layers adsorbed traces of pyridine from the laboratory atmosphere, and the pyridine interfered with the spectrophotometric evaluation of reserpine.

Solvents employed should be of high quality, e.g. MFC grade, and free from contamination. Unless the solutes are known to be very stable, low boiling point solvents are preferable, minimizing thermal decomposition during drying.

Boundary effect, a problem particularly associated with mixed solvents, must be avoided by ensuring vapour saturation in the tank. Boundary effect is caused by differential evaporation across the plate, the rate of evaporation being more rapid at the edges of the plate than in the middle. This is due to the large volume of the tank relative to the volume of the plate and to the presence of a pocket of unsaturated atmosphere behind the plate. Thus the R_f of a solute is greater near the edges of the plate and a row of solute spots across the plate will present a concave appearance. As elution technique involves the removal of a row of spots or a streak across the plate, it is desirable that this should be linear. Therefore, tanks must be lined with filter paper saturated with solvent mixture and adequate time must be allowed for equilibration of the atmosphere in the tank. Tanks should be as small as is practicable. Davies (1963) demonstrated that such conditions also favour the production of compact spots.

3. SOLUTES

Stability of the solute to be assayed is a key factor in quantitative thin-layer chromatography. In some cases qualitative separation is best performed in the dark, e.g. *Rauwolfia* alkaloids (Court, 1966), and quantitative estimation of such photosensitive substances necessitates operating in darkness and subdued light. Exposure to ultraviolet or direct visible light caused marked changes in the optical densities of reserpine in chloroformic solution at 268 nm and rescinnamine in methanolic solution at 302 nm (Harris *et al.*, 1968).

In attempting to assay complex mixtures, e.g. from plant extracts, problems may be encountered due to the variable proportions of components present. Thus Shellard (1964) advisedly stated that reference to published R_f values for individual substances can be misleading. Some substances in admixture yield a greater R_f value than that given by an equivalent load of marker substance, others yield a lesser R_f value. Variation in the loads may also cause reversal of the relative positions of the substances. For satisfactory quantitative thin-layer chromatography it is therefore essential to confirm the identity of the substance assayed, e.g. by chromogenic reactions or ultraviolet spectrophotometry.

4. SEPARATION TECHNIQUE

Provided that standardized procedures are adopted and identical apparatus and conditions are employed, reproducible results will be obtained. The extension of the development distance from the usual 10–15 cm or more may facilitate a greater linear separation of solute spots, although the R_f values may not change appreciably (Table 1). Such greater linear separation ensures a more efficient and accurate separation of the required adsorbent/solute areas. Two-dimensional thin-layer chromatography does not adapt readily for the elution technique as spots tend to become more diffuse as a result of the second development and large numbers of plates are usually necessary if a reasonable amount of solute is to be recovered for estimation.

TABLE 1

Separation of *Rauwolfia* alkaloids using solvent running distances
of 10 cm and 15 cm

Alkaloid	R_f values $\times$ 100		Linear distances in cm	
	10 cm	15 cm	10 cm	15 cm
Reserpine	33	33	3·3	4·9
Rescinnamine	31	30	3·1	4·5
Yohimbine	20	19	2·0	2·9
Ajmaline	4	3	0·4	0·5
Ajmalicine	52	51	5·2	7·6
Reserpiline	26	22	2·6	3·3
Serpentine	0	0	0·0	0·0

Mean values of 25 determinations; solvent system: *iso*-octane : acetone : *n*-butanol (58 : 33·6 : 8·4); adsorbent: Silicagel G.

B. THE APPLICATION OF THE TEST AND REFERENCE SAMPLES

Having established an effective separation system, the technique of applying test and reference materials quantitatively to the plates must be considered. Two alternatives are available:

(a) the application of a series of spots at regular intervals along the starting line, and

(b) the application of a continuous streak of test or reference solution along the starting line.

Various pipettes and syringes are commercially available, e.g. Hamilton syringe (Hamilton Co., Whittier, California, U.S.A.), Agla micrometer syringe (Burroughs Wellcome and Co., London), "Lambda" micropipettes and Microcaps disposable micropipettes (Shandon Scientific Company Ltd., London), micropipettes (Quickfit and Quartz Ltd., Stone), etc., but repetitive spotting of 1–5 μl. volumes is a tedious operation and various mechanical devices, e.g. syringes on suitable movable tables or trolleys, have been described (Ritter and Meyer, 1962; Vandenheuvel, 1966).

Streaking the test or reference solution along the baseline is not an easily performed manual technique and this too has been improved by the use of apparatus such as the trough device of Achaval and Ellefson (1966) and the automatic applicator designed by Coleman (1964).

Opinions differ concerning the advantages of spotting or streaking. Millett *et al.* (1964) preferred spotting for alkaloidal solutions and experiments in our laboratories proved more satisfactory when spotting was employed. Streaking does, however, make better use of the plates, reducing the number required.

The band width or spot diameter must be as small as possible because, as the chromatogram develops, the areas occupied by the spots tend to increase and the band width widens due to diffusion. Nybom (1967) observed

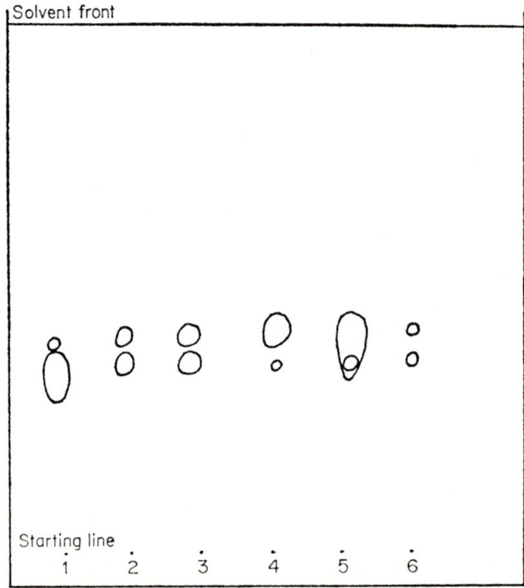

FIG. 1. The effect of load on the resolution of a mixture of reserpine and rescinnamine.

	Load (μg)					
	1	2	3	4	5	6
Upper spot	0·5	2	3	5	15	0·5
Lower spot	10	4	3	1	0·5	0·5

Solvent system: *iso*-octane : acetone : *n*-butanol (58 : 33·6 : 8·4); Silicagel G layers.

that there is no direct relationship between the amount per spot and the spot size, variations being due to the characteristics of the individual substances and to layer thickness. Nevertheless the load applied is critical and affects both the shape and position of the spot after development. Overloaded spots produce comet-like vertical streaks; if the adsorption isotherm is convex, the R_f value measured from the central point of the streak will be greater than normal and, if the adsorption isotherm is concave, the R_f value will be less than normal. The tail of the streak usually interferes with other spots on the chromatogram, rendering the plate useless for elution (Fig. 1). Therefore the load to be applied must be decided by preliminary experimentation.

The technique of applying microlitre quantities of solutions on to thin layers is not easy. Fairbairn has referred to the accuracy of methods of delivering small volumes. In practice, microlitre volumes often do not readily drop from the needle of the syringe on to the thin layer and great care must be exercised to avoid pricking the surface of the adsorbent because a hole made in this manner causes distorted triangular or crescent-shaped spots on the subsequently developed chromatogram (Truter, 1963) (Fig. 2).

If losses are to be minimized, the normal method of drying applied spots in a gentle current of air must be avoided and, therefore, low boiling point solvents are necessary to ensure compact spots.

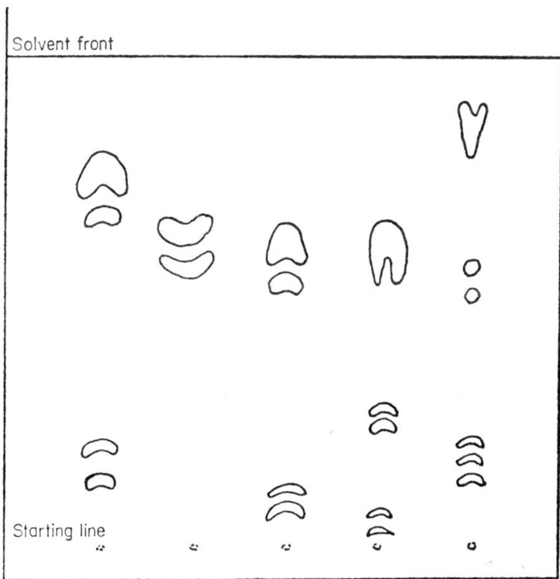

FIG. 2. Distorted shapes of spots, due to excessive pricking of the surface of the adsorbent.

C. THE RECOVERY OF THE ASSAY AND REFERENCE SAMPLES

For the elution technique most recovery methods depend on the removal of the adsorbent and adsorbed solute from the developed and dried thin-layer plate prior to the elution of the solute with an appropriate solvent. The required areas on the plate are first located either by their colour if it is distinctive, or by fluorescence in screened ultraviolet light, or by comparison with marker compounds in a detector lane on the edge of the plate. Thus, for example, the *Rauwolfia* alkaloids can be reasonably recognized by their fluorescence colours under ultraviolet light (Court, 1966) (Table 2). Alternatively, these alkaloids could be detected by spraying the detector lane with Dragendorff's reagent and the corresponding horizontal strips of adsorbent marked (Fig. 3). In this case the absence of boundary effect is essential.

TABLE 2

Recognition of *Rauwolfia* alkaloids by fluorescence colours as seen under screened ultraviolet light

Alkaloid	R_f value $\times$ 100	Fluorescence colour
Ajmalicine	48	Apple-green
Ajmaline	13	Violet
Aricine	55	Orange
Deserpidine	34	Blue-green
Rescinnamine	28	Blue-green
Reserpiline	23	Yellow
Reserpine	32	Apple-green
Reserpinine	54	Apple-green
Serpentine	0	Intense blue-violet
Yohimbine	18	Blue

Solvent system: methanol 10: methylethylketone 30: heptane 60; Silicagel G adsorbent.

The areas in which the required adsorbed compound is located can be removed by careful scraping with a spatula or scalpel and transferred to suitable flasks or beakers for further treatment. Gänshirt and Morianz (1960) employed scraping on to parchment paper or foil with subsequent brushing into the elution vessel. Such manoeuvres need to be performed carefully in the absence of air currents to prevent losses.

In order to reduce such losses various types of suction device on the vacuum cleaner principle have been recommended (Mottier and Potterat, 1955; Mottier, 1958; Morrison and Chatten, 1965; Fairbairn and El-Masry 1967). In the earlier devices cotton wool pads were employed to prevent adsorbent passing into the suction line, but later modifications used sintered glass discs, the apparatus then functioning also as an elution tube (Fig. 4).

Recently Hunt (1967) reported an improved model which obviates the accidental introduction of sebum from the operator's fingers into the eluate, and this avoids the risk of contamination of steroid eluates in particular. For large scale work Ritter and Meyer (1962) designed a suction unit which transferred the adsorbent directly into a Soxhlet thimble.

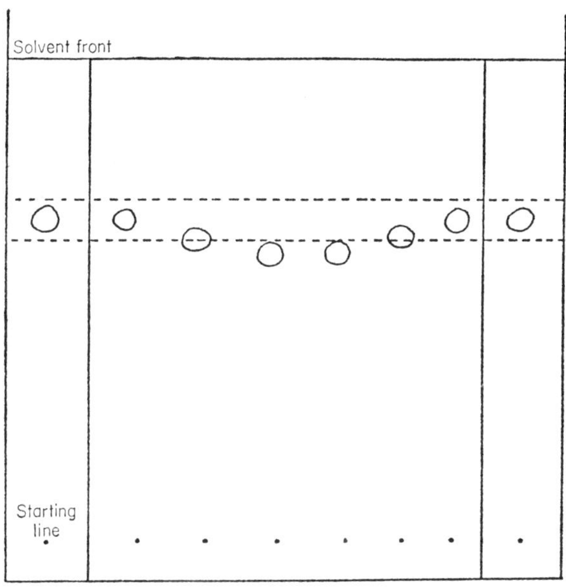

FIG. 3. Location of a compound by means of marker lanes. Boundary effect shown here must be avoided.

Elution of the solute from the adsorbent may be simple agitation with a solvent followed by removal of the adsorbent by centrifugation, or may involve column or micro-Soxhlet extraction methods. Incomplete elution is an obvious source of error and various pieces of apparatus have been suggested to improve elution efficiency. Millett *et al.* (1964) employed a small column into which solvent was introduced by a small wick leading from the solvent reservoir. A microchromatography column permitting elution under pressure was recommended by Lehmann *et al.* (1967). Attal *et al.* (1967) transferred the adsorbent to a 3 ml syringe fitted with a 25-gauge needle and a small plug of cotton wool in the bottom of the syringe; acting as a column, this apparatus proved valuable in eluting steroids and gave high recovery values.

Choice of eluting solvent is governed by the nature of the adsorbed substance and of the adsorbent. Either complete recovery or a constant percentage recovery must be attainable. Aqueous solvents are limited in value because calcium sulphate is slightly soluble in water, and polar solvents such as ethanol employed to elute hydrophilic substances also extract constituents of the thin layers. Possible elution solvents can be determined experimentally by developing test plates of the adsorbent and adsorbed substance. If the R_f value of the latter exceeds 0·8, the developing solvent will probably elute the substance successfully (Bobbitt, 1963).

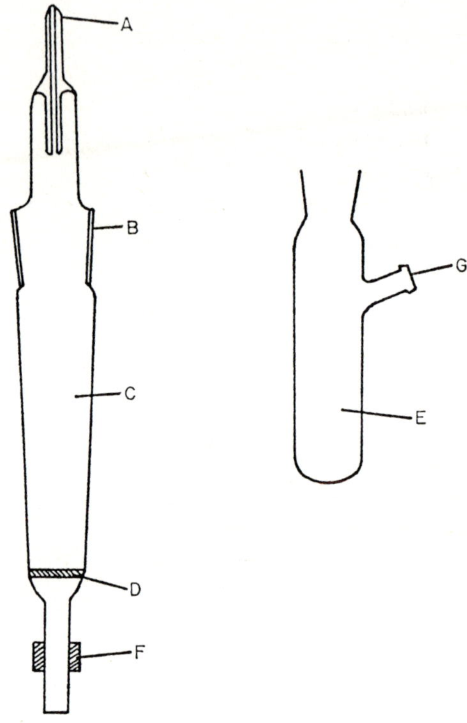

Fig. 4. Microvacuum cleaner for the removal of spots from thin-layer chromatograms. The nozzle A has a ground glass end inclined 45° to normal. The ground glass joint B enables the removal of the nozzle assembly. In operation the adsorbent is sucked into chamber C and is prevented from entering the suction line by the sintered glass filter D. After suction the nozzle assembly is replaced by a stopper, the tube then being used for elution. To remove the eluate after agitation with a solvent, tube E is attached at position F and by application of vacuum at G, the eluate can be withdrawn.

D. Measurement Methods

The microanalysis of the resultant eluate has been performed in several ways including gravimetric estimation, ultraviolet spectrophotometry, colorimetry, fluorimetry, polarography and radiometry.

The gravimetric method depends on direct weighing of the residue after evaporation of the eluting solvent. Williams *et al.* (1960) reported that gravimetric methods yielded poor results. Results tend to be high and three obvious sources of error are:

(a) Extracted substances from the adsorbent are included in the residue weighed.
(b) The total quantity of residue recovered from a few chromatoplates is, at the most, only a few milligrams and special microbalances must therefore be available.
(c) The required substance may not be totally recovered in the elution process.

The gravimetric method was used by Purdy and Truter (1963) for estimating the quantitative composition of cabbage wax, and by Dunn and Robson (1965), who determined the quantitative composition of fatty ester mixtures.

Spectrophotometric methods comprise two groups:

(i) Direct measurements in the ultraviolet or visible wavelength range, and
(ii) indirect measurements after the substance has been reacted with a suitable chromogenic reagent.

In either case the eluate is adjusted to a known volume and transferred to a suitable spectrophotometer or colorimeter. Results are calculated by reference to appropriate calibration curves or extinction coefficients.

Initial inspection of the absorption spectrum produced by the recording spectrophotometer reveals the absorption maxima of the substance. The relationship of the concentration of the substance and the optical density or absorbance at the maximal wavelengths is plotted graphically, a straight-line graph indicating that the substance obeys the Lambert-Beer law for the specific concentration range. Subsequent estimations are made within this concentration range.

An advantage of the direct, ultraviolet spectrophotometric method is that the identity of the substance assayed can be confirmed by its absorbancy peaks. Thus Schlemmer and Link (1959) refer to the absorption maxima for reserpine at 268 nm and 295 nm, the ratio of the extinction coefficients being 1·60.

Difficulties arise from the presence of substances extracted from the adsorbent, such substances absorbing electromagnetic energy and interfering with absorbancy measurements, especially in the 200–250 nm region. Harris *et al.* (1968) examined ethanolic extracts of silicagel G and observed a prominent peak at 218 nm which was a summation of peaks yielded by silicagel and gypsum. In an attempt to remove this peak, these workers extracted the silicagel by agitation for 2 hr with ethanol. After drying this adsorbent and preparing and activating thin layers from it, the plates were developed with ethanol, dried and the adsorbent scraped off prior to extraction with absolute ethanol in a Soxhlet apparatus for 3 hr. The resultant solution still yielded a main peak at 218 nm although absorption in the 260–340 nm range was considerably reduced.

The occurrence of extraneous matter even after prewashing causes high results and several workers have investigated methods of offsetting this error by the use of spectrophotometric blank solutions prepared from the chromatographic adsorbent. Gänshirt and Morianz (1960), finding that the contribution of the adsorbent to the spectrophotometric absorption could only be reduced to about one-half by prewashing, extracted an area of silicagel G equal in area to and adjacent to the test area on their plates and used this as the blank. They observed that this technique of preparing blanks considerably improved the accuracy of the assay. Harris *et al.* (1968), however, found that arbitrary removal of areas from the corresponding positions on plates developed under similar conditions to the test plates produced variable results. Better results were obtained by extracting an equivalent weight of adsorbent from a corresponding area of a developed and dried blank plate and using this extract as the spectrophotometric blank. It was noted that the optical density of the blank increased with up to 3 hr Soxhlet extraction with ethanol but prolonged extraction resulted in no further increase.

Gänshirt and Morianz (1960) proposed an alternative scheme for compensating the interference due to the adsorbent. The optical densities of the unchromatographed substance and the substance recovered from the chromatogram are recorded over a wide wavelength range, pure solvent being used as the spectroscopic blank. Baselines are constructed across the two peaks and it is assumed that the gap between the two baselines represents the contribution of the impurities to the optical density (Fig. 5). This method of correction was also employed by Pechtold and Rentsch (1967) in their method for the estimation of adenosine in pharmaceutical preparations.

Despite improved replication when such precautions are taken, the main limitation of the elution method is the incomplete recovery of the substance from the adsorbent. Thus Gänshirt and Morianz (1960), recovering the methyl and propyl esters of *p*-hydroxybenzoic acid from silicagel G, could only achieve an average extraction of 95·6% and 93·1% respectively. In our laboratories the best average recovery of reserpine from silicagel G was 97·4%.

To allow for all the systematic errors in the spectrophotometric method, Gänshirt *et al.* (1960) determined specific extinction coefficients by chromatographing known quantities of reference compounds and determining the optical densities of the resultant eluates. For this technique the average deviations from the means was 2·3%, based on ten determinations each of 50 μg samples of six different bile acids.

Another method of minimizing the incomplete recovery problem is the use of a chromogenic reagent to yield a more readily extractable complex and this usually avoids the difficulties due to the adsorbent impurities, which interfere mainly in the 200–300 nm waveband, visible coloured complexes being of longer wavelength, e.g. Gagliardi and Pokorny (1967) estimated a cobalt complex at 645 nm wavelength. The colour complex employed must obey the Lambert-Beer law within the limits of the assay, i.e. a linear relationship of colour intensity and concentration of substance. The chromophore is not necessarily specific and, therefore, the initial separation on the thin-

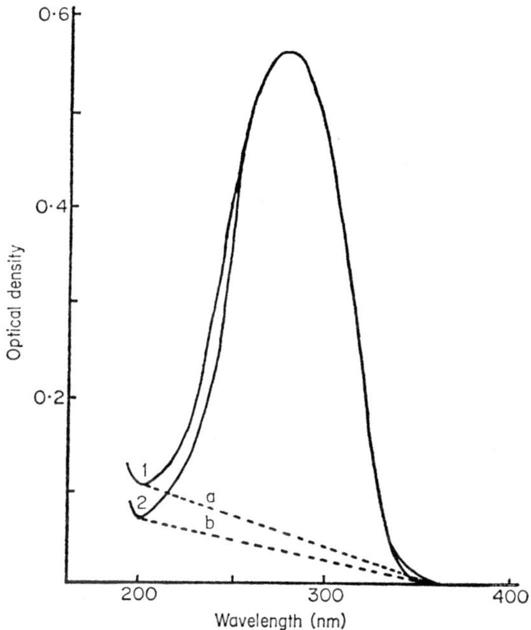

FIG. 5. Gänshirt and Morianz method for the estimation of the contribution of constituents of the adsorbent to the optical density.
Curve 1. Sample of propyl-*p*-hydroxybenzoate recovered from the chromatogram.
Curve 2. Original sample.
The contribution of the adsorbent is represented by the gap between the baselines a and b.

layer plate must be efficient. For example, several *Rauwolfia* alkaloids react with nitrous acid to yield a yellowish-green complex, which can be estimated quantitatively at 390 nm wavelength. Haycock and Mader (1957) showed that any *Rauwolfia* alkaloid possessing the A, B, C ring skeleton and 11-methoxy group of reserpine would yield the complex. Thus reserpine, rescinnamine, reserpiline and reserpinine, all possessing the 11-methoxy group, produce the complex, whilst aricine, which has a 10-methoxy group, yields a weak reaction and other *Rauwolfia* alkaloids with neither 10- or 11-methoxy groups, e.g. ajmalicine, deserpidine and yohimbine, produce no colour complex (Fig. 6).

An alternative stratagem is to remove the adsorbent from the plate and generate the colour complex *in situ* in the adsorbent which can then be removed by centrifugation. In this manner Walsh *et al.* (1965) estimated lipids by the colour reaction with hydroxamic or dichromic acids.

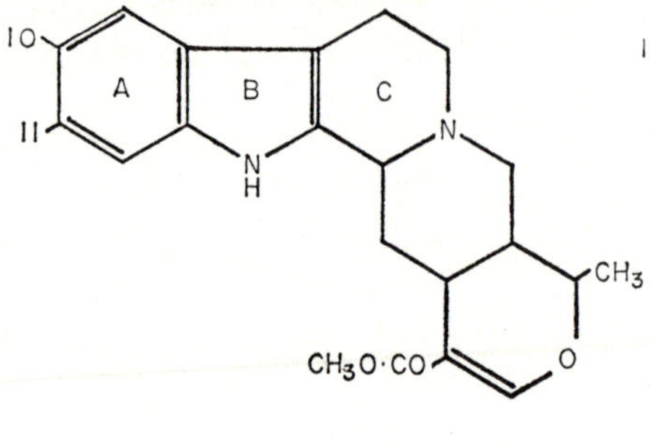

FIG. 6. Structures of *Rauwolfia* alkaloids relating to colorimetric assay based on the 11-methoxy-A,B,C ring structure.

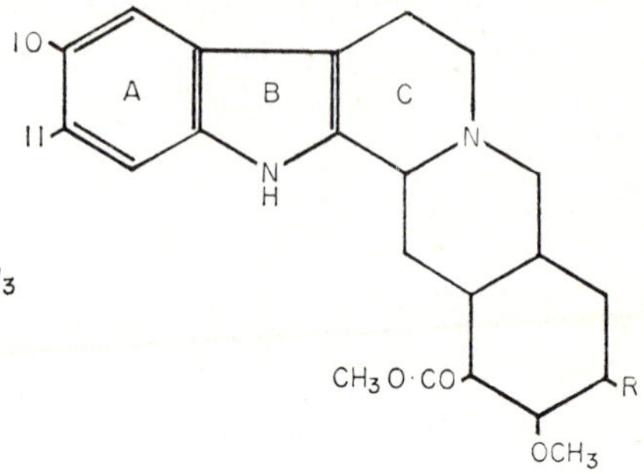

I. Tetrahydroalstonine group alkaloids.

	10	11	Colour complex
Reserpinine	H	CH_3O	Positive
Aricine	CH_3O	H	Weak
Ajmalicine	H	H	Negative
Reserpiline	CH_3O	CH_3O	Positive

II. Reserpine group alkaloids.

	10	11	R	Colour complex
Reserpine	H	CH_3O	TMB	Positive
Rescinnamine	H	CH_3O	TMC	Positive
Deserpidine	H	H	TMB	Negative

TMB = trimethoxybenzoic ester; TMC = trimethoxycinnamic ester.

In general colorimetric methods depend on a shift to longer wavelengths for determinations. Struck (1961) evolved a method for the determination of oestrogens based on the absorption maxima in alkaline solution (at 242 nm instead of the normal 280 nm). This shift to a shorter wavelength reduces the accuracy of the method despite the use of spectroscopic blanks.

For fluorescent substances, fluorimetric estimation has been suggested. Unfortunately the intensity of fluorescence is not simply related to the concentration of solute except in a small range at low concentrations, and the observed intensity of fluorescence is normally related to the concentration by means of a standard curve. Particulate matter, e.g. adsorbent particles, dust, etc., can cause considerable variation in results. The method was first applied to the estimation of 7-geranoxycoumarin by Vannier and Stanley (1958), but the accuracy of the method was not high. Subsequent workers have reported more accurate applications (Bruinvels, 1963; Rutkowska and Wojsa, 1966).

Oelschlager *et al.* (1965) determined an eluate containing 2-nitro-4-aceto-amidophenetol by a polarographic technique and Ch'en *et al.* (1966) determined strychnine in nux vomica by elution and catalytic polarography. An infrared method was proposed by McCoy and Fiebig (1965) and, in the microassay of vincamine, Karacsony *et al.* (1965) eluted the alkaloid and estimated it by non-aqueous titration.

Radioactive substances also can be separated by thin layer techniques, eluted and the intensity of the emission measured in a suitable Geiger-Muller counter. Methods have been proposed for labelled lipids (Stryder, 1964), prochlorperazine (Seno *et al.*, 1964) and steroid creams (Comer and Hartsaw, 1965).

E. ACCURACY AND APPLICATION

In assessing the merits of an assay procedure, one must consider:

    (a) the nature of the substance to be assayed;
    (b) the scientific equipment available and its sensitivity;
    (c) the technical demands on the operative, especially if manipulative techniques are involved;
    (d) the time available; and
    (e) the alternative methods available, if any, and their relative accuracies.

The elution technique is normally encountered in association with spectrophotometric, colorimetric or fluorimetric estimations (Tables 3–7), the gravimetric procedure being unreliable and yielding high results. Fluorimetric methods are summarized in Table 7. The accuracy of such methods is not generally considered to be high and, in any case, these methods can only be applied to substances fluorescing under the conditions of assay. Colorimetric methods are, likewise, restricted to substances forming coloured complexes, the optical densities of which can be measured in the spectrophotometer, provided that the Lambert-Beer law is obeyed under the conditions of assay. The advantage of these methods is that measurements made at longer wavelengths are much less affected by interference due to extracted substances from the adsorbent. By comparison of Tables 3, 5 and 7, colorimetric procedures would appear more accurate than fluorimetric methods, and as good as direct spectrophotometric assays. Direct spectrophotometric methods can yield accurate results (Table 3), although low coefficients of variation can only be obtained if meticulous care is exercised.

Four sources of error are:

    (a) inaccurate delivery of reference and test solutions on to plates (see Chapter 1);
    (b) incomplete extraction from the starting material and inefficient separation of the components of the extract on the plates, especially

where there is wide variation in the respective proportions of components;

(c) incomplete desorption of the substance to be estimated; and

(d) interference by substances extracted from the adsorbent.

Assessing various quantitative thin-layer chromatography methods, Purdy and Truter (1964) shewed that the mean coefficient of variation for published spectrophotometric assays was 5·3% (based on 200 observations) if correction for adsorbent impurities was made, and 2·3% (based on 60 observations) if calibrated extinction coefficients were employed. As the latter routine compensates for most of the variables, it is preferred although it must be stressed that accurate results demand considerable care. Purdy

TABLE 3

Spectrophotometric analyses for which coefficients of variation are available

Substance	Wavelength (nm)	Coefficient of variation (S.D. %)	Authors
Biphenyl	248	3·5	Kirchner *et al.* (1954)
Biphenyl	248	9·3	Stanley *et al.* (1957)
Reserpine	268	5·5	Schlemmer and Link (1959)
Rescinnamine	268	3·8	Schlemmer and Link (1959)
Bile acids	385–391	1·5–2·8	Gänshirt *et al.* (1960)
Methyl-*p*-hydroxybenzoate	256	3·2	Gänshirt and Morianz (1960)
Propyl-*p*-hydroxybenzoate	256	4·8	Gänshirt and Morianz (1960)
Steroids	242	7·9–10·1	Struck (1961)
2-4-Dinitrophenylhydrazone of acetaldehyde	355	8·0	Nano (1964)
Furoic acid	253	0·66–1·56	Millett *et al.* (1964)
Hydroxymethyl furoic acid	259	0·31–1·83	Millett *et al.* (1964)
Chloramphenicol	278	1·66	Kassem *et al.* (1966)
Ergot alkaloids			
Ergosine	318	2·7	Prochazka *et al.* (1966)
Ergocristine	318	0·7	Prochazka *et al.* (1966)
Ergocornine	318	1·1	Prochazka *et al.* (1966)
Ergokryptine	318	0·7	Prochazka *et al.* (1966)
E K 115	318	3·4	Prochazka *et al.* (1966)
Reserpine	268	1·32	Harris (1966)
Opium alkaloids			
Noscapine	313	0·82	Fairbairn and El-Masry (1967)
Papaverine	251	2·0	Fairbairn and El-Masry (1967)

and Truter preferred the simple elegance of spot-area methods despite the slightly greater accuracy of the spectrophotometric methods. Choulis (1967), on the other hand, preferred elution technique to spot area for the determination of adrenalin, although no statistical data was offered, and Morrison and Chatten (1965), estimating barbiturates, decided that a colorimetric method was better.

The wide range of application of the elution technique includes determinations of alkaloids, glycosides, antibiotics, steroids, metals, plant pigments, etc. (Tables 3–7), and, in particular, the method is applicable to complex

TABLE 4

Other spectrophotometric methods

(Insufficient data available for assessment of the coefficient of variation)

Substance	Authors
Squill glycosides	Streidle (1961)
Opium alkaloids	Mary and Brochmann-Hanssen (1963)
Cannabis	Korte and Sieper (1964)
Hydrocortisone	Cavina and Vicari (1964)
Ergot alkaloids	Prochazka et al. (1965)
Senna glycosides	Longo and Meinardi (1965)
Benzoates, hydroxybenzoates	Pirella et al. (1966)
Rutin	Rodriguez and Alves (1966)
Cape Aloes	Longo and Fumagalli (1966)
Steroids	Cristol and Bienfait (1966)
Ataractics, antihistamines	Sarsunova et al. (1966)
Adrenalin	Choulis (1967)
Plant pigments	Petrova (1966)
Adenosine	Pechtold and Rentsch (1967)

TABLE 5

Colorimetric analyses for which coefficients of variation are available

Substance	Coefficient of variation (%)	Authors
Neomycin	1·9	Foppiano et al. (1965)
Digitalis glycosides	2·0	Lindic and Repic (1966)
Cobalt	1·0–3·5	Gagliardi and Pokorny (1967)
Bismuth	4·5–5·2	Gagliardi and Likussar (1967)
Azorubin	3·2	Lehmann et al. (1967)
Tartrazin	1·2–1·8	Lehmann et al. (1967)

TABLE 6

Other colorimetric analyses

(Insufficient data available for assessment of the coefficient of variation)

Substance	Authors
Esters of aliphatic acids	Vioque and Holman (1962)
Pregnanediol in urine	Bang (1964)
Hydrocortisone	Cavina and Vicari (1964)
Noradrenaline	Choulis (1965)
Rhaponticin	Borkowski and Sobiczewska (1965)
Ergot alkaloids	Prochazka *et al.* (1965)
Barbiturates	Morrison and Chatten (1965)
Sorbic acid	Mueller and Braxl (1966)
Cholesterol	Horvath (1966)
Strophanthus glycosides	Lutomski *et al.* (1966)
Ergot alkaloids	Wichlinski and Skibinski (1966)
Ephedrine, codeine	Sarsunova and Chi (1966)
Antihistamines, ataractics	Sarsunova *et al.* (1966)
Dinocap (DPC)	Chiba and Yatabe (1966)
3-Methoxy-4-hydroxymandelic acid in urine	Strobach (1967)
Hyoscyamus alkaloids	Paris and Saint-Firmin (1967)
Uranium	Gagliardi and Pokorny (1967)

TABLE 7

Fluorimetric analyses

Substance	Coefficient of variation (if known) (%)	Authors
Geranoxycoumarin	—	Vannier and Stanley (1958)
Cinchona alkaloids Quinine, quinidine	—	Schellenberg (1962)
Aldosterone	3·4	Bruinvels (1963)
LSD	—	Cortivo *et al.* (1966)
Reserpine	2–6	Rutkowska and Wojsa (1966)
Buquinolate	5–9	Borfitz *et al.* (1967)
9-Acridanone	6·8–15·7	Sawicki *et al.* (1967)

mixtures which cannot easily be resolved by traditional methods. Unfortunately, as Tables 4 and 6 reveal, it is not always possible to obtain adequate details of accuracy from the published reports of applications and the plea, first made by Purdy and Truter, that data presented should include coefficients of variation and the weights of each compound must be repeated. As elution technique demands considerable manipulative skill, information of collaborative results, as given by Borfitz *et al.* (1967) for the determination of buquinolate in chicken feeds, is extremely valuable, yielding more convincing evidence of the practical application of a method.

The basic simplicity of the elution method, the challenging demand on operative skill and the possession of an analytical tool of great flexibility must ensure that this quantitative aspect of thin-layer chromatography will progress rapidly in the future in many fields and particularly in the fields of natural products and pharmaceuticals.

REFERENCES

Achaval, A. and Ellefson, R. D. (1966). *J. Lipid Res.* **7**, 329.

Attal, J., Hendeles, S. M., Engels, J. A. and Eik-Nes, K. B. (1967). *J. Chromatogr.* **27**, 167.

Bang, H. O. (1964). *J. Chromatogr.* **14**, 520.

Bobbitt, J. M. (1963). "Thin Layer Chromatography." Chapman and Hall, London.

Borfitz, H., Para, J., Stickles, J. V., Ginther, G. B. and Southworth, B. C. (1967). *J. Ass. offic. anal. Chem.* **50**, 264.

Borkowski, B. and Sobiczewska, M. (1965). *Acta Pol. pharm.* **22**, 529.

Brenner, M., Niederwieser, A., Pataki, G. and Fahmy, A. R. (1962). *Experientia* **18**, 101.

Brown, T. I. and Benjamin, J. (1964). *Anal. Chem.* **36**, 446.

Bruinvels, J. (1963). *Experientia* **19**, 551.

Cavina, G. and Vicari, C. (1964). "Thin Layer Chromatography", p. 180 (G. B. Marini-Bettòlo, ed.). Elsevier Publishing Company, Amsterdam.

Ch'en, C. C., Chang, S. L. and Tai, J. L. (1966). *Yao Hsueh Hsueh Pao* **13**, 131, through *Chem. Abstr.* **65**, 8673c.

Chiba, K. and Yatabe, H. (1966). *Noyaku Seisan Gijutsu* **14**, 14.

Choulis, N. H. (1965). *Hem. Hron.* **30**, 52.

Choulis, N. H. (1967). *J. pharm. Sci.* **56**, 196.

Coleman, M. H. (1964). *Lab. Pract.* **13**, 1200.

Comer, J. P. and Hartsaw, P. E. (1965). *J. pharm. Sci.* **54**, 524.

Cortivo, L. A. D., Broich, J. R., Dihrberg, A. and Newman, B. (1966). *Anal. Chem.* **38**, 1959.

Court, W. E. (1966). *Can. J. pharm. Sci.* **1**, 76.

Cristol, P. and Bienfait, R. (1966). *Ann. Endocr., Paris* **27**, 199.

Dallas, M. S. J. (1965). *J. Chromatogr.* **17**, 267.

Davies, B. H. (1963). *J. Chromatogr.* **10**, 518.

Dunn, E. and Robson, P. (1965). *J. Chromatogr.* **17**, 501.

Fairbairn, J. W. and El-Masry, S. (1967). *J. pharm. Pharmac.* **19**, 93S.

Foppiano, R., Brown, B. B. and Schlederer, E. (1965). *J. pharm. Sci.* **54**, 206.

Gagliardi, E. and Likussar, W. (1967). *Mikrochim. Acta* **1967**, 555.

Gagliardi, E. and Pokorny, G. (1967). *Mikrochim. Acta* **1967**, 550.

Gänshirt, H. and Morianz, K. (1960). *Archs. Pharm.* **293**, 1065.

Gänshirt, H., Koss, F. W. and Morianz, K. (1960). *Arzneimittel-Forsch.* **10**, 943.

Geiss, F., Schlitt, H. and Klose, A. (1965). *Z. analyt. Chem.* **213**, 321 and 331.

Harris, M. J. (1966). M.Sc. Thesis, Manchester University.

Harris, M. J., Stewart, A. F. and Court, W. E. (1968). *Planta med.*, **16,** 217.

Haycock, R. P. and Mader, W. J. (1957). *J. Am. pharm. Ass. Sci. Ed.*, **46,** 744.

Honegger, C. G. (1962). *Helv. chim. Acta* **45,** 1409.

Horvath, C. (1966). *J. Chromatogr.* **22,** 52.

Hunt, S. M. V. (1967). *Lab. Pract.* **16,** 601.

Karacsony, E. M., Gyenes, I. and Lorincz, C. (1965). *Acta pharm. hung.* **35,** 280.

Kassem, M. A., Kassem, A. A. and El-Nimr, A. E. M. (1966). *Pharm. Ztg., Berl.* **111,** 1792.

Kelemen, J. and Pataki, G. (1963). *Z. anal. Chem.* **195,** 81.

Kirchner, J. G., Miller, J. M. and Rice, R. G. (1954). *J. Agr. Food Chem.* **2,** 1031.

Korte, F. and Sieper, H. (1964). *J. Chromatogr.* **14,** 178.

Lehmann, G., Hahn, H. G. and Martinod, P. (1967). *Z. anal. Chem.* **227,** 81.

Lindic, K. and Repic, R. (1966). *Vestn. Slov. Kem. Drus.* **12,** 49.

Longo, R. and Fumagalli, U. (1966). *Boll. chim.-farm.* **105,** 767.

Longo, R. and Meinardi, G. (1965). *Boll. chim.-farm.* **104,** 503.

Lutomski, J., Kowalewski, Z. and Kortus, M. (1966). *Diss. Pharm. Pharmac.* **18,** 409.

McCoy, R. N. and Fiebig, E. C. (1965). *Anal. Chem.* **37,** 593.

Mary, N. Y. and Brochmann-Hanssen, E. (1963). *Lloydia* **26,** 223.

E. Merck AG, Darmstadt. (1962). "Dünnschicht-Chromatographie nach E. Stahl," pamphlet.

Millett, M. A., Moore, W. E. and Saeman, J. F. (1964). *Anal. Chem.* **36,** 491.

Morrison, J. C. and Chatten, L. G. (1965). *J. Pharm. Pharmac.* **17,** 655.

Mottier, M. (1958). *Mitt. Geb. Lebensmittelunters. u. Hyg.* **49,** 454.

Mottier, M. and Potterat, M. (1955). *Anal. chim. Acta* **13,** 46.

Mueller, K. H. and Braxl, Ch. (1966). *Arch. Pharm.* **299,** 560.

Nano, G. M. (1964). "Thin Layer Chromatography", p. 138 (G. B. Marini-Bettòlo, ed.). Elsevier Publishing Company, Amsterdam.

Nybom, N. (1967). *J. Chromatogr.* **28,** 447.

Oelschlager, H., Volke, J. and Lim, G. T. (1965). *Arch. Pharm.* **298,** 213.

Paris, R. R. and Saint-Firmin, A. (1967). *C. r. hébd. Séanc. Acad. Sci., Paris* **264,** 825.

Pataki, G. and Keller, M. J. (1963). *Helv. chim. Acta* **46,** 1054.

Pechtold, F. and Rentsch, H. (1967). *Arzneimittel-Forsch.* **17,** 57.

Petrova, T. M. (1966). *Dokl. vses. Akad. sel.'-khoz. Nauk.* **1966,** 16.

Pirella, S. J., Falsco, A. D. and Schwartzmann, G. (1966). *J. Ass. Offic. anal. Chem.* **49,** 829.

Prochazka, V., Kavka, F., Prucha, M. and Pitra, J. (1965). *Čsklá. Farm.* **14,** 154.

Prochazka, V., Kavka, F., Prucha, M. and Pitra, J. (1966). *Čsklá. Farm.* **15,** 363.

Purdy, S. J. and Truter, E. V. (1963). *Proc. R. Soc., B* **158,** 536.

Purdy, S. J. and Truter, E. V. (1964). *Lab. Pract.* **13,** 500.

Randerath, K. (1963). "Thin-Layer Chromatography." Academic Press, New York and London.

Ritter, F. J. and Meyer, G. M. (1962). *Nature, Lond.* **193,** 941.

Rodriguez, L. D. and Alves, M. A. P. (1966). *Rev. port. Farm.* **16,** 388.

Rutkowska, U. and Wojsa, K. (1966). *Herba Pol.* **12,** 101.

Sarsunova, M. and Chi, N. T. K. (1966). *Čsklá. Farm.* **15,** 474.

Sarsunova, M., Kakac, B., Schwarz, B., Maly, V. and Protiva, J. (1966). *Pharmazie* **21,** 752.

Sawicki, E., Stanley, T. W. and Elbert, W. C. (1967). *Talanta* **14,** 431.

Schellenberg, P. (1962). *Angew. Chem.* **74,** 118.

Schlemmer, F. and Link, E. (1959). *Pharm. Ztg., Berl.* **104,** 1349.

Seno, S., Kessler, W. V. and Christian, J. E. (1964). *J. pharm. Sci.* **53,** 1101.

Shellard, E. J. (1964). *Lab. Pract.* **13,** 290.

Stanley, W. L. (1959). *J. Ass. Offic. agric. Chem.* **42,** 643.

Stanley, W. L., Vannier, S. H. and Gentili, B. (1957). *J. Ass. Offic. agric. Chem.* **40,** 282.

Streidle, W. (1961). *Planta Med.* **9,** 435.

Strobach, H. (1967). *Z. klin. Chem.* **5,** 77.

Struck, H. (1961). *Mikrochim. Acta* **1961,** 634.

Stryder, F. (1964). *Anal. Biochem.* **9,** 183.

Truter, E. V. (1963). "Thin Film Chromatography." Cleaver-Hume Press Ltd., London.

Teichert, K., Mutschler, E. and Rochelmeyer, H. (1961). *Anal. Chem.* **181,** 325.

Vandenheuvel, F. A. (1966). *J. Chromatogr.* **25,** 102.

Vannier, S. H. and Stanley, W. L. (1958). *J. Ass. Offic. agric. Chem.* **41,** 432.

Vioque, E. and Holman, R. T. (1962). *J. Am. Oil Chem. Soc.* **39,** 63.

Walsh, D. E., Banasik, O. J. and Gilles, K. A. (1965). *J. Chromatogr.* **17,** 278.

Wichlinski, L. and Skibinski, Z. (1966). *Farm. Polska* **22,** 194.

Williams, J. A., Sharma, A., Morris, L. J. and Holman, R. T. (1960). *Proc. Soc. exp. Biol. Med.* **105,** 192.

Chapter 4

Quantitative Thin-Layer Chromatography using Densitometry

E. J. SHELLARD

Department of Pharmacognosy, Chelsea College of Science and Technology, University of London, London, England

INTRODUCTION

Densitometry is a method whereby the intensity of colour of a substance is measured directly on the chromatogram (see Chapter 2). The method was first used for measuring the concentration of amino acids separated on bands by means of electrophoresis and was later used on paper chromatography for the determination of amino acids, sugars and steroids.

Instruments designed for accommodating paper strips have been used for the indirect application of densitometry to TLC, the chromatogram having beem photographed or photocopied and the photograph or photocopy then cut into strips suitable for scanning. Dallas *et al.* (1964) scanned coloured substances on adsorbents spread on glass plates 20×20 cm by cutting them into strips suitable for fitting into a Joyce Loebl "Chromoscan" paper strip holder. More recently, Genest (1965) using a photovolt densitometer with TLC plate holder reported the analysis of lysergic acid type alkaloids in Morning Glory seeds but he gave no indication of the problems involved. Thomas *et al.* (1965) have reported the use of a photovolt densitomer-photometer with TLC plate holder for the determination of isomeric glycerides and they commented on the effect of R_f values on curve area-concentration relationship thus confirming observations made by Blank *et al.* (1958). Blunden *et al.* (1967) have reported the estimation of diosgenin in Dioscorea tubers using a "Vitatron" densitometer but again, little mention was made of the factors involved in obtaining reliable and reproducible results. Dallas (1968) and Shellard and Alam (1968), using the Joyce Loebl "Chromoscan" densitometer with TLC attachment have recently dealt in some detail with some of the factors involved in precision and reproducibility of results.

In densitometry two sets of parameters need to be considered in relation to precision and reproducibility. The first set are those relating to the instrument used and these can all be made constant; the second set are those relating to the chromatogram which will always remain variable.

A. PARAMETERS RELATING TO THE INSTRUMENT

The basis of the determination is that the coloured spot is scanned by a beam of light and some light energy is absorbed—the amount absorbed being related to the intensity of colour. This is measured by difference, that is, by returning the non-absorbed light to the photomultiplier which measures the intensity of the incident light beam or to a second photomultiplier which is in balance with the first photomultiplier. The difference in intensity between the incident light and the reflected or transmitted light is indicated by a curve drawn on a chart and/or by a built-in integrator. There must be a relationship between the area of the curve and/or the integrator reading and the amount of substance in the spot.

Several makes of intrument are available and with all of them the parameters have to be determined and made constant if they are not permanently fixed by the design of the instrument.

1. FILTERS

In order to achieve optimum results the coloured substance must absorb as much of the incident light as possible while the background adsorbent must absorb as little light as possible. As the substance is coloured it will chiefly be light of the complementary colours which will be absorbed and, since on most instruments the light source is a tungsten filament lamp emitting light of a very wide range of wavelengths it is necessary to use suitable filters to control the wavelength of the incident light. Consideration must be given, therefore, to the colour of the spot and the background adsorbent. Quite frequently the separated substances, though closely related, may have different colours so that a compromise in choice of filters must be made. The Mitragyna oxindole alkaloids give reddish violet to dark reddish brown colours when heated with ferric chloride and perchloric acid and Fig. 1 shows the range of integrated readings obtained with some of these alkaloids using a Joyce Loebl "Chromoscan" densitometer with different filters.

2. APERTURE

The shape and size of the spots are the main factors to be considered when selecting the most suitable aperture. Generally speaking, the smaller the aperture the higher the integrated readings but it is only by experiment that the most suitable aperture can be determined. Dallas found that a circular aperture 1 mm diameter was best though Shellard and Alam (1968) found that a slit aperture gave better results and that there was a lower coefficient of variation when the slit length was smaller than the width of the spot than when the slit length was greater than the width of the spot. Figure 2 shows the integrated readings obtained with two alkaloids using different apertures.

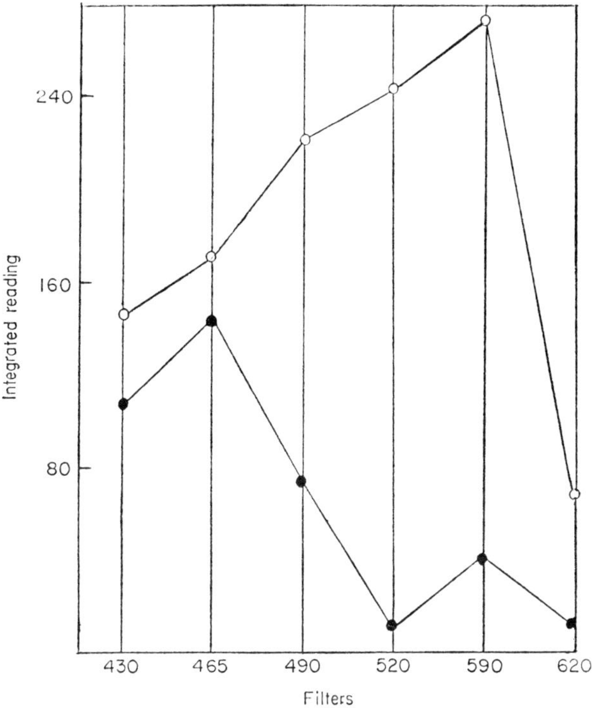

Fig. 1. Integrated readings using different filters: aperture, 0505; method of scanning, Reflectance; —○—○—, rotundifoline; —●—●—, isorotundifoline.

3. METHOD OF SCANNING

Two methods of scanning the spot are usually available: (a) reflectance and (b) transmission. Again it is only by experiment that the most suitable method for any particular determination can be decided.

The behaviour of a beam of light when it strikes an absorbing material irregularly distributed both upon the surface of and within the layer of semi-opaque particulate solid is very complex. Some consideration has been given to this problem by Shibata (1959), though his work on the spectro-photometric determination of substances in biological tissues is not strictly comparable to the absorbance of light by coloured substances adsorbed on thin layers of solid adsorbent. However, he described the optical phenomenon in any translucent material as

$$I_o = I_a + I_t + I_r + I_x$$

where I_o = incident light

I_a = absorbed light

I_t = transmitted light

I_r = reflected light

I_x = other light rays, e.g. scattered.

Thus the integrated readings and/or the curve area which are measurements of the difference between the intensity of the incident light and the non-absorbed light will depend upon the amount of non-absorbed light returned

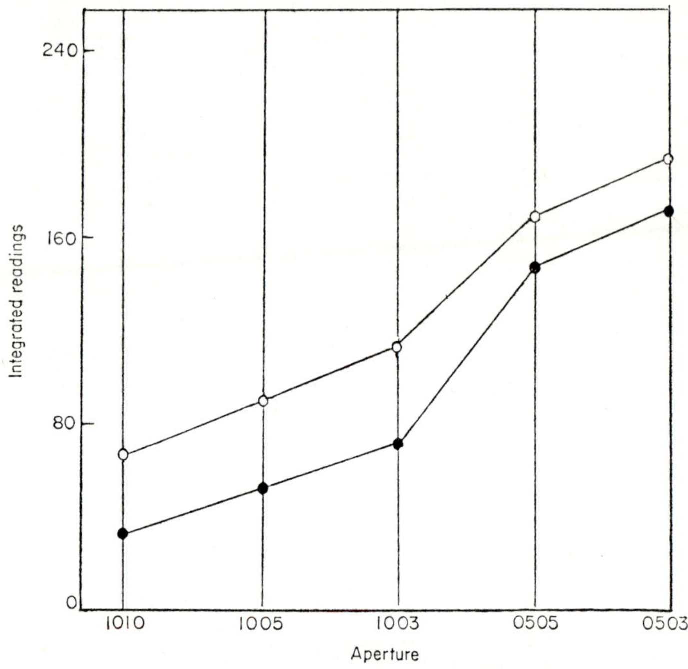

Fig. 2. Integrated readings using different apertures: filter, 465; method of scanning, Reflectance; —O—O—, rotundifoline; —●—●—, isorotundifoline.

to the photomultiplier. This in turn will depend upon the design of the instrument and upon the method used for scanning the spot. For example, with the Joyce Loebl "Chromoscan" densitometer, in reflectance measurements, only light scattered within an angle of 45° to the incident light beam passes via a mirror to the photomultiplier while in transmittance measurements no scattered light at all is received by the photomultiplier.

In the transmittance method more light will be absorbed than in the reflectance method since the beam of light in passing through the entire layer will come into contact with more of the coloured substance than when it is reflected from the surface though even in this case some penetration of the layer by the light beam will take place. It is likely too, that more light will be scattered in the transmittance method and attempts are often made to reduce this to a minimum by rendering the adsorbent itself as transparent as possible by treatment with liquid paraffin. The amount of light scattered when a spot is scanned is, however, unknown and depends on a number of factors which are not related to the design of the instrument.

4. DIRECTION OF SCANNING

The coloured spots can be scanned in two directions: (a) along the line of development of the plate and (b) perpendicular to the line of development of the plate. The choice of method will depend upon the nature of the chromatogram being examined but usually the second method is only selected when (i) the spots are not completely separated so that there is no return to the base line between two spots and (ii) impurities are deposited between the separated spots when again there would be no return to the base line.

It should perhaps be mentioned that, when scanning a number of spots either in vertical or horizontal rows care must be taken to ensure that the beam of light passes over or through the centre of the spot. This may mean some readjustments between the scanning of adjacent spots and is especially noticeable when scanning in the direction perpendicular to the line of development of the plate.

5. SPEED OF SCANNING

Some instruments provide for different rates of movement of the TL plate carrier. For any given spot, the slower the rate of scan the larger will be the integrated reading or curve area so that it is essential to maintain the same gear ratio for all observations of a particular determination.

B. PARAMETERS RELATING TO THE CHROMATOGRAM

Since all the variable factors connected with the densitometers can be fixed the most important factors in densitometry are those associated with the coloured substance at the time the colour intensity is measured. None of these can be made constant so that it is essential that the variation is kept to a minimum. Brief reference has already been made to some of the problems associated with the behaviour of a narrow beam of light when it strikes or passes through the coloured substance and it is obvious, therefore, that with fixed instrument parameters, results can only be reproduced exactly if all the coloured spots of equal weights of the same substance are identical in all respects, i.e. shape, size, profile in depth, distribution of the substance on the adsorbent (which determines the pattern of colour intensity) and its relationship with the background adsorbent.

If the substance is already coloured it is likely that the background will remain white and the measured colour intensity will show a clear relationship with the quantity of substance present. If, however, the substance is colourless it must be treated with a reagent which will produce a coloured compound contrasting sharply with the colour of the background adsorbent. The background adsorbent should preferably remain white or the colour should be so different from that of the coloured substances that, by the use of suitable filters, it can be considered as a white background.

The coloured spot should have sharp, well-defined boundaries which means that there should be little, if any, diffusion of the substance on to the surrounding adsorbent. This is unlikely to arise to any great extent during

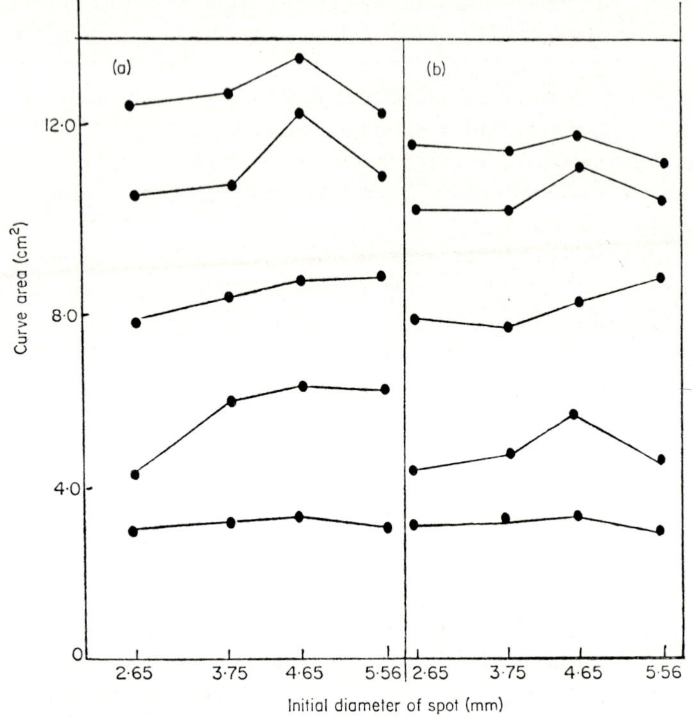

FIG. 3. Variation of curve areas with initial diameter of spot: alkaloid, rotundifoline; chromatographic system, silica gel/chloroform : acetone 514; filter, 465; gear ratio 1 : 2; aperture, 0505 for Reflectance (a); 1005 for Transmittance (b).

the separation process, but may occur as a result of spraying in order to obtain a coloured compound. The colour must be stable to light for a long time and the colour intensity, which should be at its maximum development with the particular concentration of substances must be reproducible. It is obvious, that for quantitative purposes there must be a definite relationship between the intensity of colour and the quantity of substance present. Franglen has dealt in some detail with this problem in Chapter 2.

The other factors mentioned depend upon:

(a) The shape, size and accuracy of load of the initial application.
(b) The nature of the stationary and mobile phases used in the separation process.
(c) The R_f value.
(d) The thickness of the adsorbent layer.

(e) The moisture content of the adsorbent layer.

(f) The presence of substances in the mixture other than those being examined.

(a) Fairbairn and Suwal (1959) have already drawn attention to the importance of the area of the initial application on quantitative paper chromatography while Purdy and Truter (1962) emphasized the importance of the area of the initial application on TLC where the area of the zonal spot is involved in the calculation. Thomas *et al.* (1965), Dallas (1968) and Shellard and Alam (1968) have all commented on this problem.

Dallas (1968) reported the results of three tests in which (i) 2 μl. quantities were spotted from the same capillary pipette (2 μl. capacity) as ten separate spots, (ii) 2 μl. quantities were spotted from ten different capillary pipettes (2 μl. capacity) and (iii) a comparison was made between (a) spotting five times on one spot with 2 μl. of a solution of concentration x and spotting once with 2 μl. of a similar solution of concentration $5x$. He concluded that the variation resulting from lack of precision on the initial application was greater than that arising from all the other procedures involved in densitometry put together. Fairbairn (1968) has discussed this problem in some detail in Chapter 1 and has described an automatic pipette designed by Bridger and Relph which reduces very considerably the error in applying the initial spot. Shellard and Alam, as well as investigating this problem, were also concerned with the area of the initial spot. They employed a technique similar to that described by Purdy and Truter (1962) whereby care is taken not to damage the surface of the adsorbent so that the initial applications are circular in outline.

The diameter of the applications could be varied as could the load. The results obtained are given in Table 1 and shown graphically in Fig. 3. They indicate that with the smaller loads (up to 40 μg) there is a significant difference between curve areas obtained from initial spots having fairly similar diameters but with the higher loads (over 40 μg) the difference in curve areas from initial spots having fairly similar diameters are not significant. It is interesting to note that the results obtained by reflectance and transmittance are almost the same.

(b) The most striking changes in shape and size of the separated substance results from changes in the stationary and mobile phases used for the separation process. Different systems give rise to different rates of movement, dependent on the adsorption isotherm and/or the partition coefficients of the substance with regard to the system being used. Usually a solvent system which results in slight movement of the substance produces a small round compact spot while the solvent systems which result in greater movement of the substance produces spots which are larger, frequently elongated and sometimes more diffuse laterally. These changes in shape and size are accompanied by different deposition and distribution of the substance on

Table 1(a)

Chromoscan curve area (by planimeter) of the same amount of alkaloid
but having starting spots of different diameters
Alkaloid: rotundifoline

Amount of alkaloid (μg)	Planimeter readings (sq. cm) Initial diameter of spot (in mm)			
	(a) 2·65 ± 0·35	(b) 3·75 ± 0·39	(c) 4·65 ± 0·57	(d) 5·56 ± 0·49
10·00	3·21, 2·89 2·79, 2·72 2·78, 3·12 3·14, 3·04 2·89, 2·89 = 2·953	2·89, 2·92 2·89, 3·21 3·12, 3·09 3·18, 3·07 3·00, 3·01 = 3·037	3·25, 3·15 3·16, 3·30 3·35, 3·45 3·51, 3·02 2·98, 2·99 = 3·216	2·79, 3·17 3·00, 2·80 3·09, 3·01 3·12, 2·89 2·79, 2·79 =2·945
20·00	4·05, 4·20 4·07, 4·45 3·85, 4·12 4·09, 4·23 4·75, 4·65 = 4·246	5·85, 6·05 5·85, 6·15 5·65, 5·65 5·45, 6·48 6·05, 6·05 = 5·923	6·05, 6·34 6·45, 6·85 6·07, 5·95 5·90, 5·97 6·47, 6·02 = 6·207	6·00, 6·05 5·69, 5·65 6·25, 6·17 5·85, 5·87 6·07, 6·47 = 6·007
40·00	8·12, 7·20 7·45, 8·25 8·17, 8·24 7·47, 7·12 7·49, 7·24 = 7·675	8·65, 8·55 8·34, 8·75 8·72, 8·25 8·42, 7·62 7·62, 7·74 = 8·266	9·42, 8·45 8·12, 8·25 8·19, 8·10 9·05, 9·15 9·42, 8·15 = 8·630	8·90, 9·05 8·95, 7·95 7·92, 8·15 9·15, 9·07 9·12, 9·45 = 8·77
60·00	10·70, 9·62 11·00, 10·69 10·07, 9·69 10·89, 10·69 10·71, 10·61 = 10·467	10·45, 9·96 10·23, 11·03 10·86, 10·96 10·92, 10·92 10·72, 10·62 = 10·671	12·01, 11·92 12·12, 11·76 13·00, 12·92 12·61, 12·62 12·42, 12·72 = 12·410	10·46, 10·96 11·09, 10·72 10·92, 10·62 11·21, 11·21 10·70, 10·62 = 10·851
80·00	12·01, 13·12 11·89, 12·81 12·91, 11·91 13·02, 13·20 12·81, 12·75 = 12·664	12·89, 13·21 12·85, 12·75 11·92, 12·82 12·72, 13·28 13·19, 12·82 = 12·846	14·09, 13·25 14·12, 14·02 13·75, 14·12 13·28, 13·29 13·40, 13·09 = 13·640	12·25, 12·40 12·21, 12·91 12·91, 11·82 12·45, 11·92 13·09, 11·98 = 12·390

Chromatographic system: Silica gel/chloroform : acetone (5 : 4)
Filter:　　　　　　　　465
Aperture:　　　　　　 0505
Gear ratio:　　　　　　1 : 2
Method of scanning:　 Reflectance

TABLE 1(b)

Chromoscan curve (by planimeter) of the same amount of alkaloid
but having starting spots of different diameters
Alkaloid: rotundifoline

Amount of alkaloid (μg)	Planimeter readings (sq. cm) Initial diameter of spot (in mm)			
	(a) 2·65 ± 0·35	(b) 3·75 ± 0·39	(c) 4·65 ± 0·57	(d) 5·56 ± 0·49
10.00	2·91, 2·75 2·65, 2·89 2·89, 3·01 2·91, 3·05 3·01, 3·07 = 2·914	2·95, 3·09 3·12, 3·12 2·89, 2·98 2·71, 3·09 3·10, 3·12 = 3·017	3·20, 3·31 3·09, 2·97 2·98, 3·10 3·39, 3·17 3·31, 3·20 = 3·172	2·98, 2·75 2·65, 2·72 2·65, 2·59 2·92, 2·91 2·91, 2·98 = 2·806
20·00	3·95, 4·30 4·25, 4·40 4·50, 4·49 4·55, 4·72 3·72, 3·90 = 4·278	4·35, 4·42 4·12, 4·60 4·50, 4·47 4·30, 4·80 4·95, 4·85 = 4·536	5·65, 5·45 5·70, 5·65 5·20, 6·12 5·45, 5·23 5·45, 5·23 = 5·513	3·85, 4·20 4·70, 4·05 4·35, 4·20 4·39, 4·80 4·95, 4·99 = 4·448
40·00	8·12, 7·20 7·42, 8·40 8·50, 8·65 7·21, 7·19 7·00, 7·19 = 7·688	7·92, 6·85 7·12, 7·45 7·65, 7·92 7·67, 7·21 7·85, 7·65 = 7·529	8·75, 8·62 8·72, 8·20 8·07, 7·95 7·69, 8·01 8·12, 8·13 = 8·126	8·90, 9·05 10·00, 8·03 8·92, 8·75 8·25, 8·15 8·01, 8·07 = 8·613
60·00	10·62, 10·52 10·21, 9·96 9·96, 9·71 10·21, 10·91 10·12, 10·71 = 10·293	10·71, 10·69 10·71, 9·96 9·78, 9·89 9·89, 10·02 10·21, 10·81 = 10·267	10·69, 10·91 10·71, 11·09 11·12, 11·92 11·12, 11·91 11·21, 11·90 = 11·258	9·97, 9·81 9·71, 10·21 10·21, 10·81 10·92, 10·98 10·79, 10·92 = 10·433
80·00	11·28, 11·81 10·81, 11·99 12·01, 11·98 11·56, 11·62 11·52, 11·91 = 11·649	11·21, 10·92 11·91, 12·21 12·01, 12·17 12·09, 10·89 11·29, 10·91 = 11·561	10·89, 12·08 12·12, 12·08 11·98, 11·81 11·91, 11·91 11·81, 11·98 = 11·857	10·91, 10·98 11·81, 10·92 11·81, 11·61 11·71, 10·81 10·91, 10·89 = 11·236

Chromatographic system: Silica gel/chloroform : acetone (5 : 4)
Filter: 465
Aperture: 0505
Gear ratio: 1 : 2
Method of scanning: Transmittance

the adsorbent. Thus, the same quantity of the same substance, will give, after separation by TLC using different solvent systems, different responses with regard to the densitometer readings. Figures 4 and 5 show the effect of different solvent systems on the calibration curve of rotundifoline.

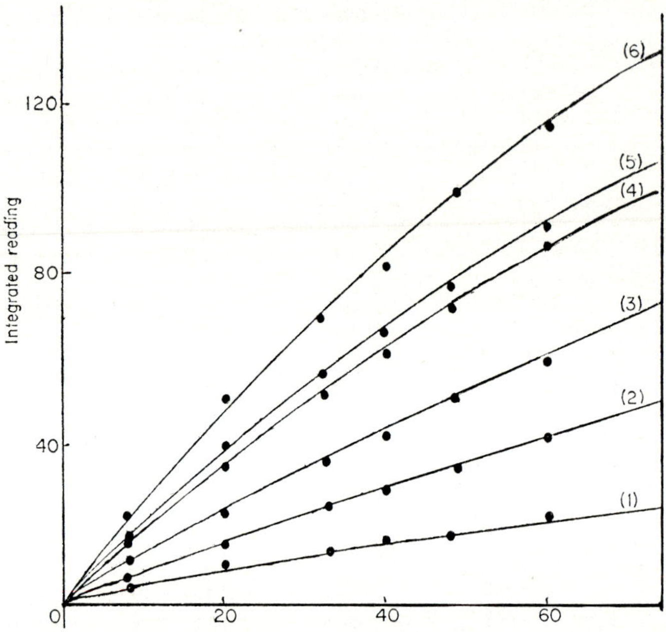

FIG. 4. Variation of calibration curve with different solvent systems: alkaloid, rotundifoline; filter, 465; aperture, 1005; gear ratio, 1 : 2; method of scanning, Reflectance; solvent system, (1) chloroform : cyclohexane 7 : 3, (2) chloroform, (3) benzene : ethyl acetate 7 : 2, (4) ether, (5) benzene : acetone 1 : 1, (6) chloroform : methanol 95 : 5.

(c) R_f values can also vary within one system as a result of variations of a number of factors, e.g. distance travelled by solvent, rate of movement of solvent, etc. Shellard (1964) and Dallas (1965), Geiss *et al.* (1965) and others have discussed the many factors which effect reproducibility of R_f values.

The distance usually selected in TLC is 10 cm (on 20×30 cm or 20×5 cm plates) but this may be increased to 12 or 15 cm to achieve a better separation or it may be reduced to 8 cm on precoated foils. Dallas (1968) claims that the curve area (even at constant R_f values) is affected by the rate of flow of the solvent and states that the higher the rate of flow, the larger the curve area. This is more marked at higher R_f values than at low R_f values. He also investigated the effect of the time the substance was in contact with the solvent on the curve area but he concluded that the effect, in normal practice, is negligible.

Figures 6 and 7 show the relationship between curve area and R_f values (for reflectance and transmittance) and it will be observed that for both

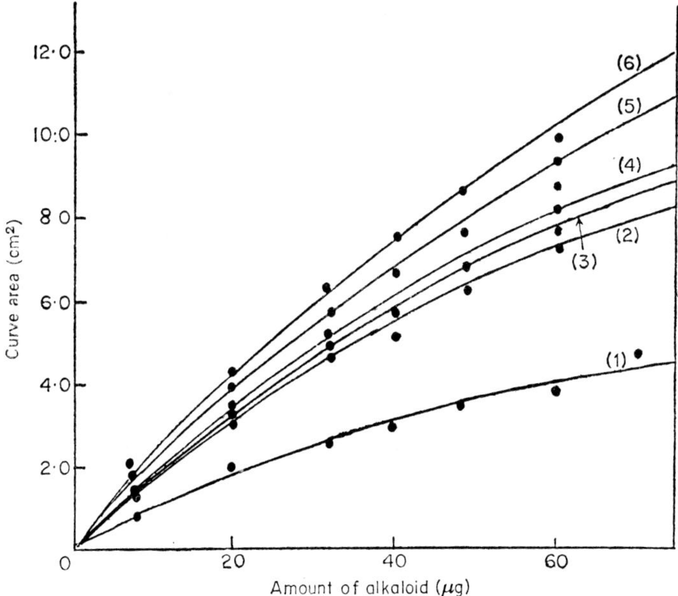

FIG. 5. Variation of calibration curve with different solvent systems: alkaloid, rotundifoline; filter, 465; aperture, 1005; gear ratio, 1 : 2; method of scanning, Transmittance; solvent system, (1) chloroform : cyclohexane 7 : 3, (2) chloroform, (3) benzene : ethyl acetate 7 : 2, (4) ether, (5) benzene : acetone 1 : 1, (6) chloroform : methanol 95 : 5.

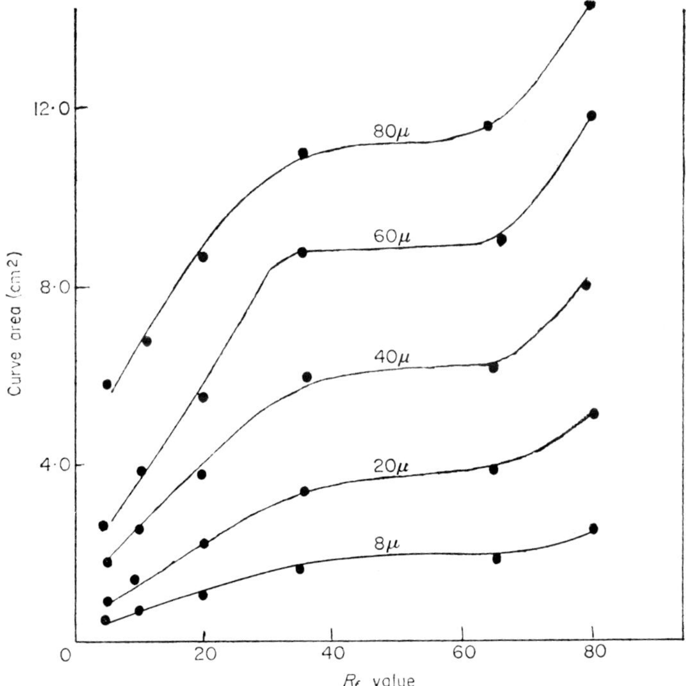

FIG. 6. Variation of curve area with R_f value: alkaloid, rotundifoline; filter, 465; aperture, 1005; gear ratio, 1 : 2; solvent system, chloroform : acetone 5 : 4; method of scanning, Reflectance.

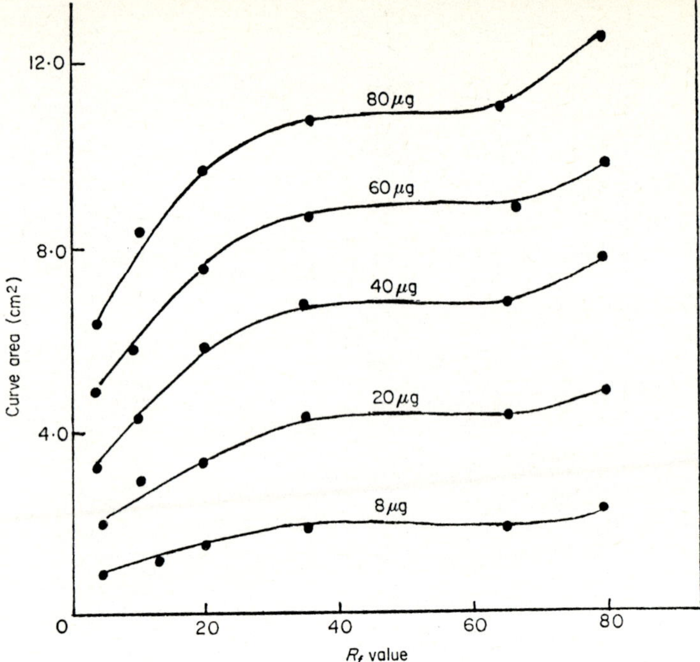

FIG. 7. Variation of curve area with R_f value: alkaloid, rotundifoline; filter, 465; aperture, 1005; gear ratio, 1 : 2; solvent system, chloroform : acetone 5 : 4; method of scanning, Transmittance.

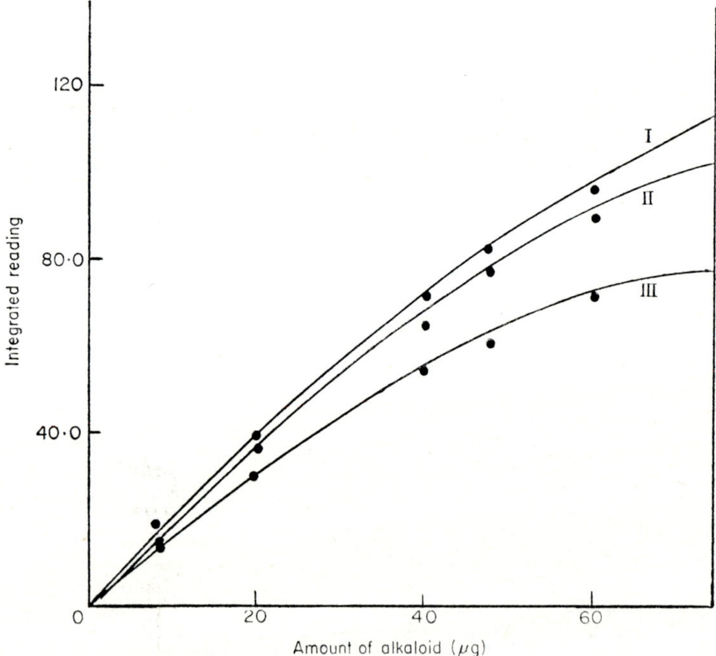

FIG. 8. Variation of calibration curve with distance travelled by solvent: alkaloid, rotundifoline; filter, 465; aperture, 1008; gear ratio 1 : 2; solvent system, chloroform : acetone 5 : 4; method of scanning, Reflectance; distance travelled by solvent, (I) 15 cm; (II) 10 cm; (III) 5 cm.

methods of scanning there is an increase in the densitometer readings with increase in R_f value irrespective of the amount of alkaloid though for R_f values between 0·35 and 0·65 there is less variation, and the readings may be regarded as fairly constant. This is an agreement with the results of Thomas *et al.* (1965) who reported that the areas of the curves varied with R_f but were fairly constant within the range 0·3–0·8.

Figure 8 shows the variation in densitometer readings with different distances moved by the solvent front.

(*d*) R_f values may be affected by variation in thickness of the layers but even with identical R_f values the curve area depends upon the thickness of the layer. Dallas (1968) investigated this by using uniformly graded thicknesses of silica gel and developing the plates in an "S" tank in which R_f values are independent at layer thickness (Honegger, 1963). Some of his results are given in Table 2.

It will be seen that with Reflectance the curve area decreases as the thickness of the layer increases while with Transmittance the curve area increases as the thickness of the layer increases.

Little is known about the actual distribution of the coloured substance upon the particles of adsorbent making up the layer but it is clear that the thickness of the layer must affect the profile in depth of the coloured substance

TABLE 2

Variation of curve area with layer thickness (Dallas)

Plate no.	Thickness (μ)	R_f	Curve area (sq. cm)	
			Reflectance	Transmittance
1	130	0·31	13·13	15·08
2	300	0·31	11·38	18·50
3	390	0·31	10·54	20·53
4	470	0·31	10·01	22·05
5	520	0·31	9·66	22·31

and its distribution within the layer. From the results it appears that more light is absorbed when the beam of light passes through thicker layers but that less light is absorbed when the beam of light is reflected from the surface of the thicker layers.

If the substance was evenly distributed within a cylinder of adsorbent the reduction in absorbance would be in inverse ratio to the thickness of the layer when the Reflectance method is used (assuming no penetration of light past the surface of the layer) but with the Transmittance method there would be no change in the amount of light absorbed whatever the thickness of the layer. However, it is extremely unlikely that the coloured substance is distributed evenly throughout a cylindrical portion of the layer and it is probable that there is a greater concentration nearer the upper surface of the layer (Fig. 9). In surface view there is a concentration in the central

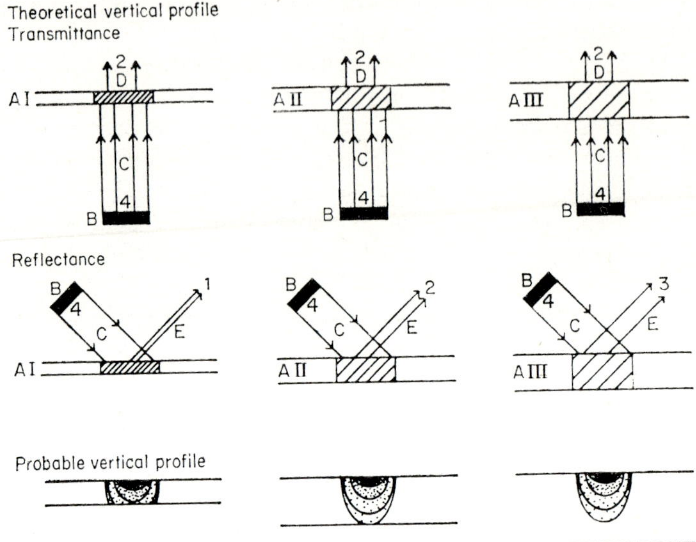

FIG. 9. Vertical profiles of developed solute on thin layers: A. layer of adsorbent, thickness A III 3 × A I; thickness A II 3 × A I; B. aperture; C. intensity of incident light; D. relative intensity of transmitted light; E. relative intensity of reflected light.

region elongated in the direction of development with the distribution falling off evenly towards each margin. There is, however, a greater concentration towards the front of the spot than towards the rear as indicated by the shape of the curves obtained by scanning the spots by reflectance both along the line of development and perpendicular to it (Fig. 10).

(e) The moisture content of the adsorbent may vary to such an extent that the resultant R_f values may also vary considerably. In view of the importance of reproducible R_f values in obtaining reproducible curve areas or integrator readings it is necessary to stress that only layered plates which have been stored in carefully controlled humidities after spreading should be used. However, since adsorbents readily take up moisture depending upon the relative humidity and temperature at which they are stored (Dallas, 1965; Geiss *et al.*, 1965). Dallas (1968) investigated the effect of moisture content on the curve areas irrespective of variation due to R_f value. He found that a change of 3% in relative humidity led to a change in the order of 1% in the curve area and he suggested that this might be due either to a change on the translucence of the adsorbent or to a change in the adsorption spectrum because of changes in the strength or nature of the adsorption forces.

(f) The shape and size of the spot after separation will often depend upon the presence of other substances in the mixture and the extent to which they can be completely separated. In their examination of Mitragyna oxindole

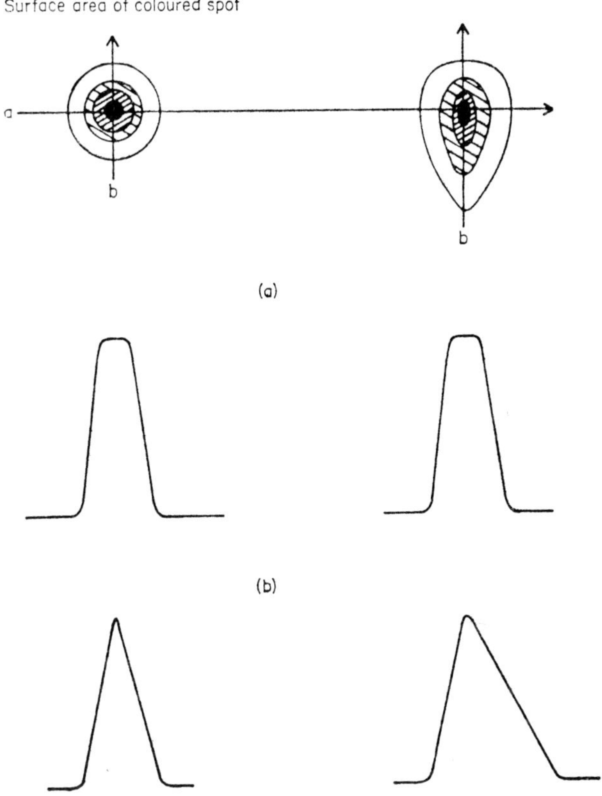

FIG. 10. Horizontal profiles of developed solution on thin layers with appropriate scanograms. (a) Curves obtained by scanning perpendicular to line if development; (b) curves obtained by scanning along line of development.

alkaloids, Shellard and Alam found that with artificial mixtures of four alkaloids the coefficient of variation was about 5, but that in examining plant material containing 2, 3, 5, 7 and 11 alkaloids the results indicated that when only two or three alkaloids were present the coefficient of variation was less than 4, when 7 or 11 alkaloids were present the coefficient of variation was up to 10 and for one particular alkaloid (iso-rotundifoline) it was 15·92. The crowding of the alkaloids caused distortion of the shape of some of them. Some alkaloids were so close to each other that overlapping curves were obtained and it was necessary to extrapolate geometrically in order to complete the curve and measure their areas.

C. CALCULATION OF RESULTS

Two methods are available for determining the amount of substance present on a chromatogram.

(1) If only one substance is involved which can be separated easily from other substances in the mixture and if the approximate amount of substance present is known, it is possible to apply to the same plate, as alternate initial applications, varying loads of the pure substance along with the substance under examination. By measuring the curve areas of all the chromatographed spots it is easy to ascertain the spot of known amount which is identical with that of the unknown. It may be desirable to construct a calibration curve from the data obtained from the known quantities and then to assess the amount present in the unknown. Blunden *et al.* (1967) using a "Vitatron" instrument employed this procedure for the estimation of diosgenin, establishing the relationship

$$\log_{10} \text{weight} \propto \sqrt{\text{integrator reading}}$$

They stated that their experimental error was approximately 7%.

(2) When several substances are involved it is necessary, after establishing a relationship between the amount of substance and the curve area or integrator reading, to construct beforehand, calibration curves for each substance with each TLC system likely to be used for the separation of the substances. Because of the many variations in the spots to be scanned, some of which have already been discussed, the calibration curves must be constructed as a result of a number of separations and curve area measurements. There are, however, additional variations which must be taken into account. These are concerned with the precision and reproducibility of the curve area. Both Dallas (1968) and Shellard and Alam (1968) have investigated these variations as far as the Joyce Loebl "Chromoscan" densitometer is concerned.

The optical system of this instrument is so designed that when there is a balance between the reference beam of light and the light reflected from the background adsorbent the pen records a continuous straight line and the integrator is not counting. When light is absorbed by the coloured spot, the pen describes a curve, the peak of which corresponds to maximum absorbance. At the same time the integrator counts, the rate at which it counts increasing as the pen moves away from the base line. If the margin of the coloured spot is sharp and if the adsorbent in between the adjacent spots is white then the pen returns to the same base line every time, the integrator stops counting and the area of the curve should be related linearly to the integrator count.

Three questions arise: (i) will the same spot when scanned several times give the same curve area each time? (ii) will the same spot when scanned several times give the same integrator count each time? and (iii) is there a truly linear relationship between curve area and integrator count?

Dallas (1968) scanned a single spot five times by reflectance, removing the plate and repositioning it after each scan. Each recorded curve area was thus measured ten times by planimetry and he calculated the percentage deviation for the 50 readings at 1·73. Shellard and Alam (1968) obtained similar results with spots of different loads and repeated the tests using the integrator in which case the percentage standard deviation was 2·68. They showed that under these conditions there is a linear relationship between the integrator count and curve area.

Unfortunately this ideal situation is rarely obtained. The adsorbent between the spots is not always free from absorbing substances with the result that the pen does not always return to the previous base line but creates a new, higher base line. (This can be overcome to some extent by scanning perpendicular to the direction of development.) The area of curve arising from the new base line will give a higher integrator count than the same area of curve will give from the original base line and in this situation the relationship between curve area and integrator count is no longer linear (Fig. 11). For this reason it is preferable to use curve area measurements rather than integrator counts. Measurement of curve areas can be achieved by the usual methods but there is less variation of results if a good planimeter is used, although it may take longer.

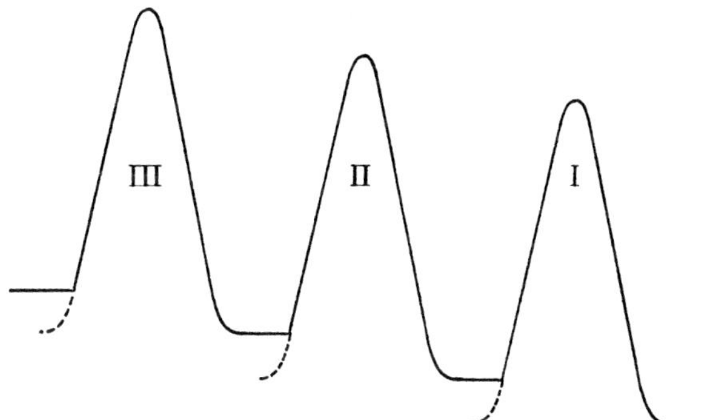

FIG. 11. Area A, $A_I = A_{II} = A_{III}$; Integrator count C, $C_I < C_{II} < C_{III}$.

Tables 3(a) and (b) show some results obtained on the same chromatograms using curve area and integrator count.

The variations associated with the actual measurement of curve area in addition to those arising from variation in the spot itself make it necessary to take the average curve area of a number of determinations when calculating the results of unknown concentrations.

In their work with the Mitragyna oxindole alkaloids Shellard and Alam converted a curvo-linear relationship between curve area and amount of alkaloid into a straightline relationship by plotting curve area $\propto$ square

Table 3(a)

Variation in the areas of the chromoscan curves (by planimeter) of the
same concentration per spot due to different base lines
Alkaloid: isomitraphylline

Amount of alkaloid (μg)	Area of curves (sq. cm)				
	Distance of base line from left-hand side of Chromoscan chart				
	30·00 mm	35·00 mm	40·00 mm	45·00 mm	50·00 mm
80·00	15·25	17·65	17·45	17·50	17·40
60·00	14·25	16·52	16·59	16·70	16·74
40·00	11·15	12·52	12·69	12·50	12·72
20·00	7·01	8·07	8·12	8·05	8·05
12·00	4·02	4·79	4·82	4·85	4·75
8·00	2·07	2·87	2·89	2·92	2·85

Chromatographic system: Silica gel/chloroform : acetone (5 : 4)
Filter: 465
Aperture: 0505
Gear ratio: 1 : 2
Method of scanning: Reflectance

Table 3(b)

Variation in the Chromoscan Integration Readings of the same
concentration due to different base lines
Alkaloid: isomitraphylline

Amount of alkaloid (μg)	Integrated area readings				
	Distance of base line from left-hand side of Chromoscan chart				
	30·00 mm	35·00 mm	40·00 mm	45·00 mm	50·00 mm
80·00	178	210	215	255	250
60·00	159	198	205	210	230
40·00	129	156	167	176	221
20·00	81	96	107	126	137
12·00	45	59	68	76	98
8·00	29	37	36	39	54

Chromatographic system: Silica gel/chloroform : acetone (5 : 4)
Filter: 465
Aperture: 0505
Gear ratio: 1 : 2
Method of scanning: Reflectance

root of weight of alkaloid. This was confirmed by a statistical analysis of the results. However, the fact that there was a considerable variation of the curve area for the same amount of alkaloid both on the same plate and, more particularly from plate to plate, led them to utilize the regression line equation for each alkaloid as a means of determining the amount of alkaloid present in unknown mixtures.

The regression line equation is:

$$x = b(y - c)$$

where x = amount of substance present (μg)

b = slope of regression line

y = grand mean from several plates *or* average of a number of readings from one plate

c = intercept

By using a number of readings from a number of plates (Shellard and Alam used five readings from each of six plates) and determining the grand mean rather than taking the average of a number of readings, say 15, from one plate, more reliable results can be obtained. This method takes into account the probable variations between one plate and another (moisture content, thickness, etc.) and although the curve area readings from any one plate show greater reproducibility than the readings between different plates, the final result may be far from correct.

Table 4 shows results obtained for isorotundifoline using both methods. With method (b) six different results were obtained varying from 93% to 107% of the correct answer.

TABLE 4

Determination of isorotundifoline by two methods. (a) Average of 30 readings (five readings from each of six plates). (b) Average of 15 readings from one plate (six determinations)

Amount of alkaloid (μg)	Method	Amount of alkaloid found (μg)	Coefficient of variation
16·0	(a)	16·04	4·52
	(b)	16·97, 15·31, 14·86 16·79, 15·86, 17·14	2·46–2·82
		average = 16·16	5·63

SUMMARY

Densitometry is a quick and reliable method for determining the amount of substance directly from the chromatogram after its separation from a mixture. However, when used with thin-layer chromatography, it is necessary to take special precautions in preparing the chromatogram and developing the colour complex since the reproducibility and precision of the results depend almost entirely upon the actual nature and distribution of the colour complex at the time it is scanned by the densitometer. Because of the variation in different chromatograms it is better to calculate the result from a regression line equation using the mean from a number of determinations on several plates rather than from a calibration curve and the data from one plate.

ACKNOWLEDGEMENT

I would like to thank Dr M. Z. Alam for carrying out much of the practical work discussed in this chapter.

REFERENCES

Blank, M. J., Schmidt, J. A. and Privett, O. S. (1958). *J. Am. Oil chem. Soc.* **41,** 371.
Blunden, G., Hardman, R. and Morrison, J. C. (1967). *J. Pharm. Sci.* **56,** 948.
Dallas, M. S. J. (1965). *J. Chromatogr.* **17,** 267.
Dallas, M. S. J. (1968). *J. Chromatogr.* **33,** 193 and 337.
Dallas, M. S. J., Barret, C. B. and Padley, F. P. (1964). *Joyce Loebl Publication,* 3.
Fairbairn, J. W. and Suwal, P. N. N., Jr. (1959). *Pharm. acta, helv.* **34,** 561.
Geiss, F., Schlitt, H. and Klose, A. (1965). *Z. anal. Chemie* **213,** 321 and 331.
Genest, K. (1965). *J. Chromatogr.* **18,** 531.
Honegger, C. G. (1963). *Helv. chim. Acta* **46,** 1772.
Neubauer, D. and Mothes, K. (1961). *Planta Medica* **9,** 466.
Purdy, S. J. and Truter, E. V. (1962). *Analyst* **87,** 802.
Shellard, E. J. (1964). *Lab. Pract.* **13,** 290.
Shellard, E. J. and Alam, M. Z. (1968). *J. Chromatogr.* **33,** 347.
Shibata, K. (1959). *Meth. Biochem. Anal.* **7,** 77.
Squibb, R. L. (1907). *Nature, Lond.* **198,** 317.
Thomas, A. E., Scharoun, J. E. and Ralston, H. (1965). *J. Am. Oil chem. Soc.* **42,** 790.

Chapter 5

Quantitative Thin-Layer Chromatography using Fluorimetry

D. E. JÄNCHEN

Camag, Muttenz, Switzerland

Since thin-layer chromatography has been generally accepted as an analytical separation method there is a steadily increasing demand for quantitative techniques. Methods including the removal of the separated compounds from the plate are time-consuming and painstaking. Therefore, the trend in quantitative TLC is towards *in situ* methods.

Apart from the so-called "spot area measurement", which is based entirely on the apparent size of the spot disregarding its concentration, there are four "optical" methods for quantitative *in situ* scanning:

(1) The transmission or densitometric methods,
(2) The reflection or remission technique,
(3) Measurements of the fluorescence, and
(4) The fluorescence quenching method.

The term "fluorimetry" comprises the two last-mentioned techniques. For a better understanding the expressions "fluorescence" for method (3) and "quenching", i.e. fluorescence quenching, for method (4) shall be used.

Comparing the four different techniques one with the other gives the following picture: Of the four only number (3), measuring the fluorescence, is a direct method. Here, the higher the quantity of the compound to be determined, the higher is the intensity of the emitted light. Therefore, within the limits of the instrument, it is possible to adjust the sensitivity of the measurement to the requirements. This is a definite advantage over the three indirect methods, because there the sensitivity is determined by the background of the plate.

In fluorescence measurements where the background appears dark to the instrument, most of the sources of error occurring in indirect methods are eliminated, such sources of error being: ambiguous relationship between sample concentration and light attenuation, spot deformation, and light scattering (Klaus, 1964).

One argument often used in favour of densitometry or remission measurements is versatility. However, in the majority of cases the compounds are colourless and, therefore, have to be reacted with a suitable reagent before they can be scanned. But the application of such reagents introduces another source of error. Actually there seem to exist more substances with native fluorescence than there are compounds absorbing light in the optimum range for photometric measurements, i.e. between 400 and 700 nm, at least if one disregards the group of dyestuffs.

As far as versatility is concerned, fluorimetry is not disfavoured if one takes into account: (a) fluorogenic reagents, and (b) the fluorescence quenching method.

Generally, the following procedures for fluorimetric *in situ* scanning of TLC plates can be applied:

CF, chromatography (followed by) fluorescence measurement (the compound shows native fluorescence);

CRF, chromatography, reaction, fluorescence measurement (the separated, non-fluorescent compound must be rendered fluorescent by a spray reagent);

RCF, reaction, chromatography, fluorescence measurement (the compound is rendered fluorescent prior to chromatography, whereby the reaction can be carried out before or after sample application);

CQ, chromatography, quenching method;

CRQ, chromatography, reaction, quenching method;

RCQ, reaction, chromatography, quenching method.

Although there was interest in fluorimetric scanning of TLC plates no instrument capable of this type of measurement was on the market and it was decided to modify a Turner fluorometer model 111, which is a double beam instrument based on an optical bridge system. Details can be seen from Fig. 1. The decision to base the TLC scanner on this instrument was

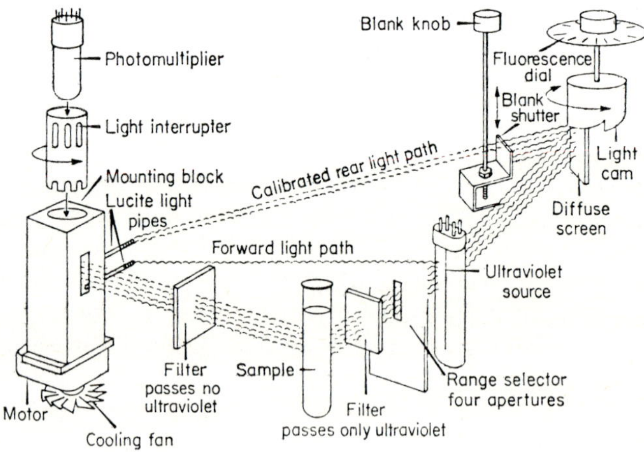

FIG. 1. Optical bridge system of the Turner Fluorometer.

made because of its high sensitivity combined with excellent stability. As the original measuring compartment of the fluorimeter was by far too small to accommodate a standard size TLC plate, the optical light path had to be changed according to Fig. 2. The primary (excitation) light passes a range selector, then the primary filter. An aperture plate very close to the surface of the layer allows to adjust the excitation slit in width between zero and 3·5 mm. The height of the slit is 15 mm and fixed. The secondary light emitted from either the spot or the layer is fed into the photomultiplier after passing the secondary filter. An S-shaped light pipe serves for feeding maximum light intensity into the photomultiplier.

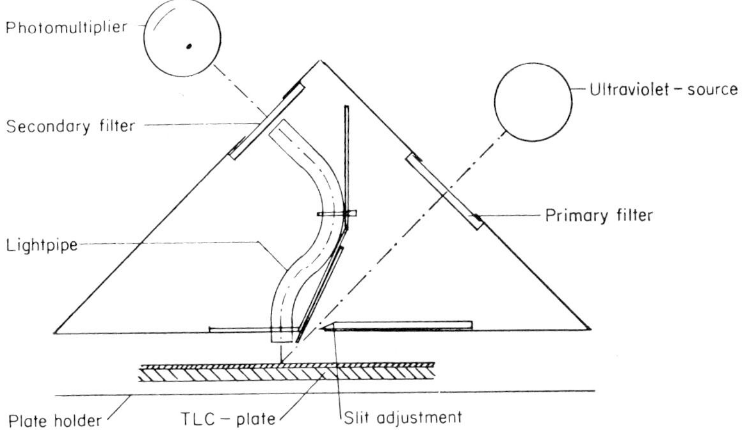

FIG. 2. Light path changed TLC scanning.

The TLC plate accommodated in the motor driven plate holder is slid at constant speed past the scanning window. The vertical position of the plate is adjusted by the rack on which the plate holder rides. The direction of scan, either parallel or perpendicular to chromatography is determined by the way the plate is inserted in the plate holder. The scanner is depicted in Fig. 3.

Low pressure mercury lamps of only 4 watts are used as light sources. Several lamps, some of them with fluorescent envelopes, allow selection of the primary light wave length between 254 and 560 nm. In general, however, only the 366 nm emission is used for all fluorescence measurements, and the 254 nm line for the quenching method. Primary and secondary filters are arranged in the light path. Their function will be mentioned later. A range selector on the primary side varies the excitation light intensity in the ratio 1 : 30. For TLC scanning in the microgram range very often the lowest intensity is still too high so that neutral density filters must be interposed on the secondary side.

Table 1 gives a survey on a number of fluorimetric determinations. The selection was made to give examples for all scanning methods as mentioned. The only exception is technique CRQ which seems to be particularly useful in lipid chromatography.

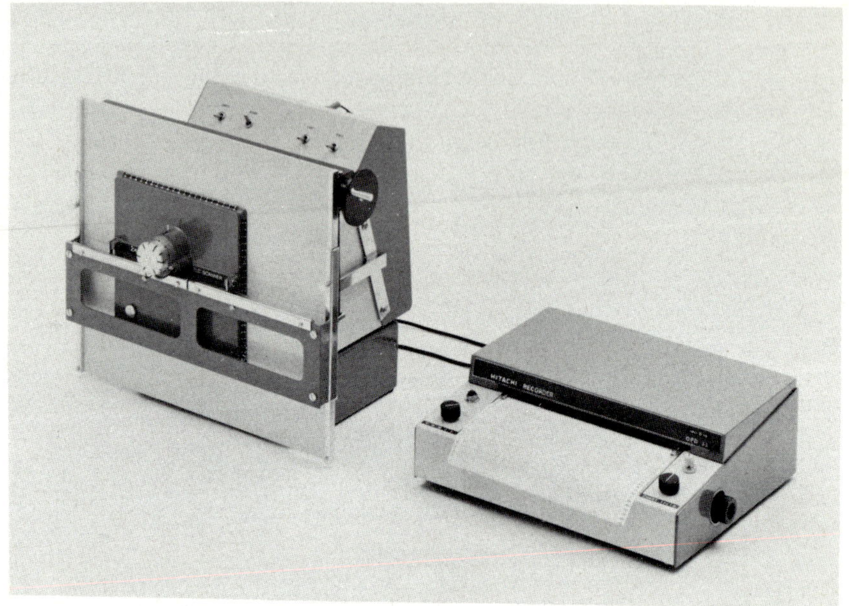

FIG. 3. Camag-Turner TLC Scanner.

From Table 1 it becomes apparent that reproducibilities of the fluorescence measurements are better than those of the quenching scans. Even a spray reagent applied in combination with the fluorescence technique affects reproducibility comparatively little.

The standard deviations reported, except example No. 1, were obtained from six samples chromatographed on the same plate. They include all errors occurring during the entire analysis. In one case, the pipetting error alone was investigated and found to be 1·4%.

The chromatographic data of the five examples can be seen from Table 2. It should be noted that in example 2 the native fluorescence of the modified cellulose formed the background for the quenching method. No "ultraviolet-indicator" was incorporated in the layer material as is usually necessary in the quenching technique.

As far as the instrument and its adjustment are concerned there are several factors influencing the scanning results:

TABLE 1

Details of fluorimetric scanning

No.	Compounds	Quantity	Symbol for scanning method	Excitation wavelength	Fluorimeter sensitivity		S.D. (%)
					Range	Transmittance of nd filter (%)	
1	Harmane alkaloids	100–300 ng	CF	366	10 ×	100	1·9
2	Nucleotides	1–5 μg	CQ	254	10 ×	40	5–7
3	Pregnanediol	0·5–5 μg	CRF	366			5
4	DNS-amino acids	1–5 μg	RCF	366	10 ×	5	3·5–5
5	PTH-amino acids	1–5 μg	RCQ	254	1 ×	20	5–7

TABLE 2

Details of chromatographic procedures

No.	Compounds	Layer	Solvent system	Spray reagent	Symbol for scanning method
1	Harmane alkaloids	Silica	Benzene-chloroform-ether 4 : 1 : 1	None	CF
2	Nucleotides	PEI cellulose	0·1 → 2 M LiCl (gradient)	None	CQ
3	Pregnanediol	Silica	Chloroform-acetone 9 : 1	Sulphuric acid	CRF
4	DNS-amino acids	Silica	Benzene-pyridine-acetic acid 80 : 20 : 2	None	RCF
5	PTH-amino acids	Silica F	Chloroform-formic acid 100 : 5	None	RCQ

References for examples: (1) Messerschmidt, 1968; (2) Pataki and Niederwieser, 1967; (3) Jänchen and Pataki, 1968; (4) Pataki and Strasky, 1966; (5) Pataki and Wang, to be published.

Choice of lamp and primary filter: Although the Turner instrument offers a larger variety of lamps including fluorescent envelope lamps with continua up to 560 nm, hitherto only two lamps have been used for TLC scanning: the general ultraviolet lamp with peak emission at 366 nm for the fluorescence method, and the far ultraviolet lamp with the main mercury line at 254 nm for the quenching technique.

The appropriate filters are: for 366 nm excitation Corning 7-60 or, if fluorescence is very bright, Corning 7–37. The Corning 7–54 filter serves for the shortwave ultraviolet-light.

Up to now no mention has been made that the scanner was used for remission measurements in the visible range. In this case the shortwave ultraviolet lamp which has a fairly intensive 436 nm emission should be used in combination with a Kodak-Wratten 2A filter (sharp cut at 415 nm).

Selection of secondary filters: About 20 sharp-cut and narrow-pass filters are available to discriminate the compound to be estimated from the background and from interfering spots. Since it is impossible to give detailed recipes for a proper filter selection, only a few basic hints will be mentioned.

The secondary filter in fluorimetry has to be non-transmittant for the excitation light. In general, excitation wavelengths are shorter than 415 nm, and therefore, the 415 nm sharp-cut filter is the one most often used for both the fluorescence and the quenching methods. It is obvious that the method can be made more selective for a certain compound emitting secondary light at a longer wavelength, when the filter either has its peak transmittance at that wavelength, or has its cut slightly below that value. However, these filters should be used only in cases where selectivity makes it necessary, because they often affect the base line of the scan. It should be mentioned that interference filters can also be inserted if so desired. Sometimes a peak is observed which starts and ends with negative deflection of the recorder. This behaviour indicates that the emission peak of the compound coincides with the border line of the filter transmittance. A typical example are optical brighteners fluorescing around 400 nm. They can be scanned using a shortwave ultraviolet lamp in combination with a secondary filter transparent below 400 nm. In such a case the user might observe fluorescence quenching under the ultraviolet lamp, provided an F-grade adsorbent was used, while the fluorimeter records fluorescence, of course, of a range invisible to the eye.

Slit-opening and scanning speed: The height of the excitation slit is 15 mm. The width of the slit may be varied between zero and 3·5 mm. If the spots on the plate are in sufficient distance from each other the slit aperture at the instrument should be set to fully open, which contributes to a uniform base line. Resolution of the scan, however, is increased when the slit is adjusted to only 1 mm. The instrument used for all five examples had a fixed scanning speed of 20 mm/min. The TLC scanner shown in Fig. 3 is equipped for dual speed, 10 and 20 mm/min. At low speed, the resolution of spots is increased similar to adjusting the slit to a smaller width, however, without affecting the uniform base line.

Direction of scan: If the chromatographic patterns permit, the scanning should be directed perpendicular to chromatography. This will give a better base line, partiuclarly with the quenching technique. If, in this case, scanning in the direction of chromatography cannot be avoided, it sometimes becomes necessary to pick up the true base line by a scan parallel to the track of the spots. Such non-uniform backgrounds occur more often with multi-component solvents. Sometimes they are noticeable by the eye under shortwave ultraviolet-light.

Sometimes there is a discrepancy in size between two neighbouring spots: In order to obtain a sufficiently large peak of the smaller spot a concentration of the sample solution has to be chosen so that the bigger spot or even both are cut into by the scanning track. In such a case, sample application in the form of a band and scanning in the direction of chromatography is the answer. The last-mentioned factor, the direction of scan, is pertinent partly to the instrument adjustment and partly to the chromatographic manipulations. Two more factors deriving from the chromatographic procedure shall be discussed:

Vapour phase saturation during chromatography: When measuring the fluorescence of a solution there is a linear relationship between fluorescence and concentration over several decades, provided the concentration was not too high. In fluorescence measurements on the TLC plate this is not the case. The calibration curves are more or less bent. This is an influence of sample concentration towards the surface of the layer, whereby this migration is caused by solvent evaporation during the chromatographic run. In order to obtain more linear calibration curves a solvent vapour phase saturation, as perfect as possible, should be attempted.

This, of course, is not only desirable in fluorimetry. Vapour phase saturation is an important facator in all "surface" methods of *in situ* scanning, but fluorescence measurements offer the best conditions for a linear relationship.

Influence of "time": By the example of certain amino acid derivatives an influence of time on the fluorescence has been reported in the literature (Pataki and Strasky, 1966; Seiler *et al.*, 1963; Seiler and Wiechmann, 1966). This influence which can be attributed to the moisture content of the layer can be eliminated either by standardizing the delay between chromatography and scanning, or by suitable reagents.

SUMMARY

The methodical advantages of fluorimetry as a method for quantitative *in situ* scanning of TLC plates are discussed. The design of a suitable instrument is described and typical results are demonstrated by some examples applying variations of the method. Some factors important for fluorimetric TLC scanning are discussed.

REFERENCES

Jänchen, D. and Pataki, G. (1968). *J. Chromatogr.* **33**, 391.
Klaus, R. (1964). *J. Chromatogr.* **16**, 311.
Messerschmidt, W. (1968). *J. Chromatogr.* **33**, 551.

Pataki, G. and Niederwieser, A. (1967). *J. Chromatogr.* **29,** 133.
Pataki, G. and Strasky, E. (1966). *Chimia* **20,** 361.
Pataki, G. and Wang, K. T., to be published.
Seiler, N., Werner, G. and Wiechmann, M. (1963). *Naturwiss.* **50,** 643.
Seiler, N. and Wiechmann, M. (1966). *Z. analyt. Chem.* **220,** 109.

Chapter 6

Advantages and Problems of Direct Spectrophotometry on Thin-Layer Chromatograms

H. JORK

*Department of Pharmacognosy and Analytical Phytochemistry,
University of the Saarland, Saarbrücken, German Federal Republic*

INTRODUCTION

Thin-layer chromatography, like paper and gas chromatography is a universally recognized method of separation, which, in combination with microanalytical techniques is responsible for many advances in modern methods of analysis.

The quantitative determination of chromatographically separated substances can be carried out in two ways. Either, the substances are removed from the adsorbent and then determined or, the quantitative determination is undertaken directly on the layer. The latter method of evaluation is normally far more sensitive than the indirect method in which elution of the substances from the adsorbent is the first part of the process (Gänshirt, 1967).

The direct methods are based upon examination of a single chromatographic spot and with TLC this is usually in the microgram or nanogram range. Direct methods involve either, the determination of radioactive labelled substances or photometric methods and these may be divided into two groups (a) densitometric and (b) spectrophotometric methods.

Densitometric methods depend upon recorded information about the differences in optical density or grey value between a chromatographic spot or zone and the surrounding adsorbent. By placing the light source and the photomultiplier on opposite sides of the plate the amount of light which passes through the layer is measured and related to the intensity of the incident light—this is the transmittance or transmission method. For reflectance or remission measurements the lamp and the photomultiplier are placed at an appropriate angle to each other and the amount of light reflected from the chromatogram is measured and related to the intensity of the incident light. Problems associated with densitometric determination have been dealt with by Shellard (Chapter 4).

In both densitometric methods only the visible range of the spectrum is used, various filters being employed to select suitable ranges of radiation. Consequently all colourless substances have to be made visible before they can be examined and this can be done either by direct carbonization with sulphuric acid or by treatment with special reagents. In many cases, however, it has not been clearly established whether the reactions are quantitative or are, at least, reproducible or even whether the reactions are dependant on concentration. This problem has been discussed in some detail by Franglen (Chapter 2).

The spectrophotometric processes operate at the wavelengths of maximum absorption by the particular substance. Thus it is possible to measure the electronic spectrum directly. Such direct evaluation by spectrophotometry possesses advantages for both qualitative and quantitative analysis compared with other methods of determination, e.g. the greater sensitivity of the measurements permits the use of ranges which, up to the present, have not been usable or only usable with some difficulty. The possibility of making all wavelengths from 200 nm upwards freely available for the evaluation of any substance including those which are fluorescent, requires an instrument which is suitable for all purposes.

A THE CHROMATOGRAM SPECTROPHOTOMETER

For the purposes of direct spectrophotometry of Thin-Layer Chromatograms, C. Zeiss, in conjunction with Stahl and Jork, have designed the Chromatogram Spectrophotometer (Fig. 1). It consists of the Spectrophotometer PMQ II, with a reflectance attachment and a mechanical stage. In order to obtain automatic recording as well as a rational qualitative and quantitative evaluation, a recording device—recorder or integrator—can be connected to the Spectrophotometer. An automatic adjustment of slit-width achieves a reduction of time in taking the spectra. A detailed descrip-

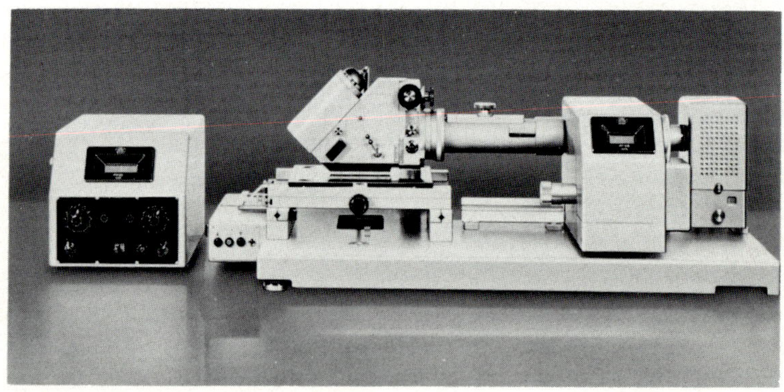

FIG. 1. The Chromatogram Spectrophotometer. (Photo: C. Zeiss, Oberkochen. F.G.R.)

tion of the prototype and of the instruments is available (Jork, 1964; Zeiss, 1967).

It is designed to measure either the varying adsorption of constantly irradiated energy or the actual emission from the substance. Thus, the evaluation of the separated spots or zones on a TLC plate can be made by measurement of transmittance or of reflectance and, in the case of fluorescent substances, by the measurement of the fluorescence. As in TLC densitometry, the degree of reflectance from the chromatogram spot is determined in relation to the reflectance from the adsorbent on the plate. Thus, the greater the amount of substance on the spot, the smaller is the degree of reflectance. In order to eliminate the interference of external influences, standard substances are developed with the samples under test on the same plates. The graphic representation of the relation between the quantity of substance and the degree of reflectance produces suitable calibration curves though in order to obtain a straight line graph passing through zero, the Kubelka-Munk theory must be applied.

Kubelka-Munk function

$$F(R_\infty) = \frac{(1-R_\infty)^2}{2R_\infty} = c \cdot \varepsilon \cdot \frac{1}{s}$$

$$c = \frac{s}{\varepsilon} \cdot \frac{(1-R_\infty)^2}{2R_\infty}$$

where R_∞ = absolute reflectance
c = amount of chromatographed sample
ε = molar extinction coefficient
s = scattering coefficient

The coefficient of variation of such measurement is ± 3 (Jork, 1968). The method of measurement by reflectance has been thoroughly investigated by Kortüm *et al.* (1963) and is used extensively in the ceramics, textile and paper processing industries for measurement of colour.

B. REFLECTANCE MEASUREMENTS

The TLC plate is placed horizontally on the mechanical stage which can be moved uniformly in the direction of the ordinates with the help of a servomotor. This is necessary for the purpose of localizing the spots (which are normally invisible) on the plate. Light of a preselected wavelength coming from the monochromator, strikes the sorbent layer vertically and the circular aperture or the slit is sharply focussed on to the plate by a system of lenses. The radiation penetrates into the layer which is partially absorbed by the chromatographed substance. Radiation which is not absorbed but diffusely reflected, reaching the photomultiplier at an angle of 45°, is measured. In this arrangement the carrier plates play no part so that they can be made of glass, metal or plastic material such as that used in the manufacture of precoated foils. Differences in thickness of the layer are much less important in reflectance than in transmittance measurements (Jork, 1968).

C. Transmittance Measurements

For purposes of transmittance measurements, the photomultiplier must be removed from the reflectance head and fixed on the optical axis directly below the mechanical stage. It is a convenient method for the examination of transparent-materials such as photographic negatives of TLC, paper-chromatograms or electropherograms.

D. Fluorescent Measurements

If fluorescent phenomena are observed, the arrangements described above for the determination of reflected radiation cannot be used because the longer wave fluorescence—mostly in the visible range of the spectrum—will reach the photomultiplier unchecked and thus lead to erroneous results. In these cases, the path of light is reversed so that the photomultiplier is behind the monochromator which then functions as a selective filter (see also Klaus, 1964) and allows only the radiation it is desired to measure, to pass through. In comparison with the Turner Fluorometer (see Chapter 5), the Chromatogram Spectrophotometer is suitable for either fluorescence and/or absorption measurements. Excitation and fluorescence spectra may therefore be taken separately as previously indicated. Similar experiments were done by Seiler *et al.* (1963) and Sawicki *et al.* (1964a, b, 1965). This applies equally to qualitative and quantitative determinations. The evaluation of fluorescent sorption layers can be carried out with the apparatus set up in almost the same manner.

The intensive irradiation of the chromatogram spots with the whole of the energy emitted by the lamp in use may prove disadvantageous since it might lead to change of labile compounds although precise measurements have in many cases shown that decomposition takes place much more slowly than the time required for the measurement (Bohrmann, 1967; Stahl and Jork, 1968; Struck *et al.*, 1968). Further, photochemical oxidation can be prevented by the application of nitrogen or some other inert gas to the reflectance head.

E. Application of Substances to the TLC Plates

The accuracy of the results obtained in quantitative Thin-Layer Chromatography depends to a considerable extent upon the precision with which the substance is applied to the plates and, the smaller the quantity of substance to be determined the more difficult it is to obtain exact reproducibility of loading (Gorbach and Haack, 1956). Only when the standard deviation of the volume measurement is small it is possible to obtain reproducibility of the results.

The substances are normally applied to the plate by means of pipettes with a volume of 5–10 μl. The accuracy and precision of some types of pipette are shown in Fig. 2. With 10 μl. pipettes the deviation from the theoretical volume is, in the case of all pipettes tested, between 3 and 5%. This systematic error occurs in the measurements both of the standard solution and the unknown and thus plays only a subordinate part in the calculations. A more important factor is the casual error and in the author's experience this is between 0·5 and 1·5%.

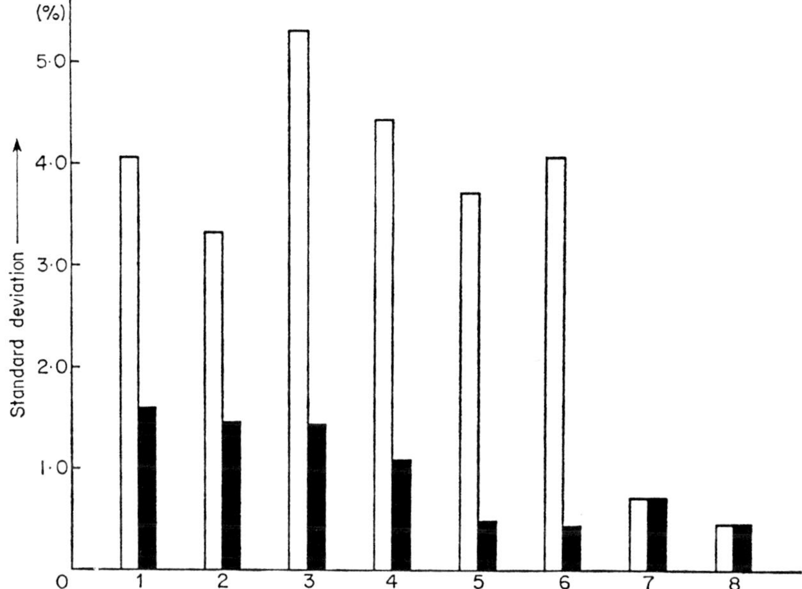

FIG. 2. Results of evaluation of various types of pipettes for accuracy and precision. ☐: Accuracy; ■: precision. **Pipettes:** 1. Microcaps; 2. Camlab capillary pipette; 3. Graduated blood pipette; 4. Lambda constriction pipette. **Microsyringes:** 5. Hamilton microsyringe; 6. Beckman submicrosyringe; 7. Desaga micrometer syringe; 8. Syringe, designed by the author.

For quantitative purposes micrometer syringes have proved most suitable. The graduations on the barrel only give rough guidance to the volume delivered and it is necessary to calibrate the micrometer scale to give precise volumes. The accuracy of the dosage is then assured by the calibration. The degree of accuracy is, however, still dependent on the amount of casual error introduced, but this may be kept to a minimum if the following points are observed:

(1) By using a short needle cemented into the syringe so that the dead space is small. The needle should have a specially hardened point and be

tapered at an angle of about 25°. The liquid will not then be taken up on the outside of the needle by adhesion.

(2) A special teflon seal is used to ensure that none of the solution passes up between the plunger and the barrel of the syringe.

(3) The syringe button consists of a flattened iron disc which is magnetic; this ensures the best possible contact with the stem of the micrometer screw and also prevents any accidental sinking of the plunger.

(4) The barrel of the syringe should be held in the hand as little as possible to prevent any increase in temperature.

With such a micro-syringe the dosage can be carried out with the utmost precision, the reproducibility being better than $\pm 0.7\%$ (Fig. 3).

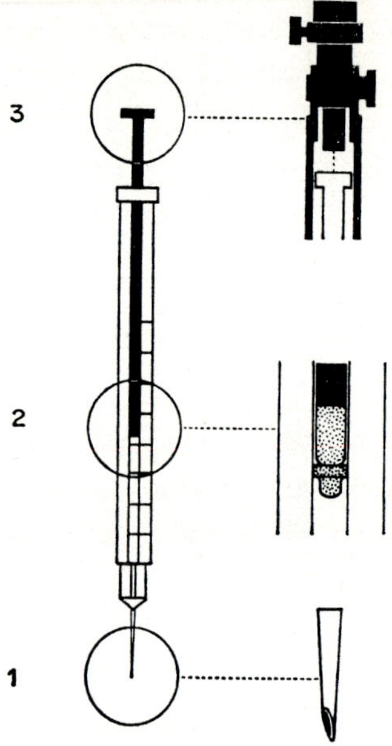

Fig. 3. Characteristics of the Jork micrometer syringe. 1. Short needle with tapered point; 2. Teflon seal; 3. Flattened magnetic syringe button.

F. Qualitative Determinations

Both absorption and fluorescence curves may be taken directly from the chromatogram spot. Figure 4(a) shows the excitation spectrum obtained with 3·1 μg Rivanol and this corresponds exactly to the number and location of the peaks in the spectrum obtained by measurement from the solution in a cuvette. Only a slight hypsochromic shift is observed for the α peak. This is stated to be due to the adsorbent (Kortüm *et al.*, 1963, 1966; Jork, 1966; Frodyma *et al.* (1964); and Frodyma and Lieu, 1967). Figure 4(b) shows the fluorescence spectrum—the dotted curve is that corresponding to the adsorbent used. The limit of detection of Rivanol by fluorescence measured is 3 ng.

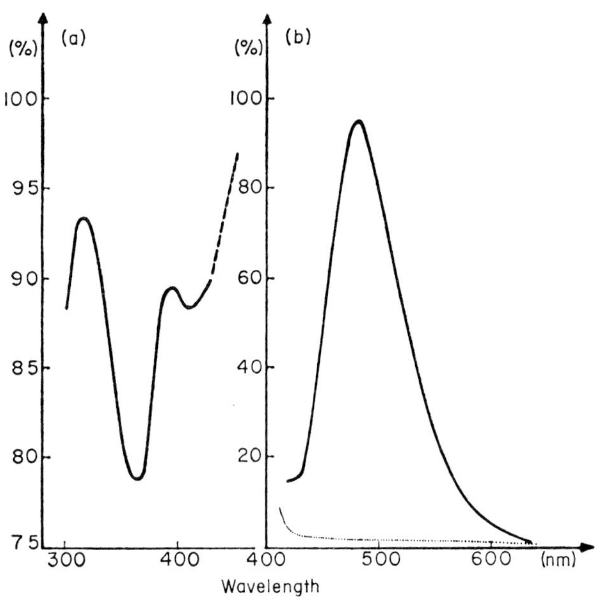

FIG. 4. (a) Absorption spectrum of Rivanol. (b) Fluorescence spectrum of Rivanol. (3·1 μg Rivanol measured directly from Silica gel H, thickness 250 μm.)

Figure 5 shows the reflectance spectrum of arbutin and for purposes of comparison, the adsorption spectrum obtained from a solution of arbutin in a cuvette is shown as a dotted line. The spot contained 1·3 μg arbutin but for the spectrum obtained from the solution 28 μg was dissolved in 10 ml. methanol. It will be seen that the reflectance spectrum shows also a slight shift of 2–3 nm compared with the absorption spectrum. This shift is also due to the effect of the medium and in the case of compounds containing carbonyl groups it may amount to as much as 8 nm. The characteristic of the spectrum is, however, preserved so that a permanent identification can be made of compounds separated on the thin layers (Stahl, 1964; Godin *et al.*, 1967). Such spectra provide more reliable information than R_f values or colour reactions; in addition, they differentiate between stereoisomeric compounds which frequently cannot be differentiated by chromatography alone.

The method is very valuable in work on "chemical races" or medicinal plants (Stahl and Jork, 1966). Eugenol methylether and its two isomers cannot be separated on a silica gel layer; their reflectance spectra, however,

have different characteristics and thus it was possible to ascertain that the
methyl ether of trans-isoeugenol is one of the main constituents of certain
essential oils of wild ginger (Stahl and Jork, 1966). Similar investigations
have been carried out with other groups of substances in carrot (Stahl, 1964)
celery (Bohrmann, 1967), fennel (Jork, 1968, unpublished) and peony root
(Pfeifle, 1967).

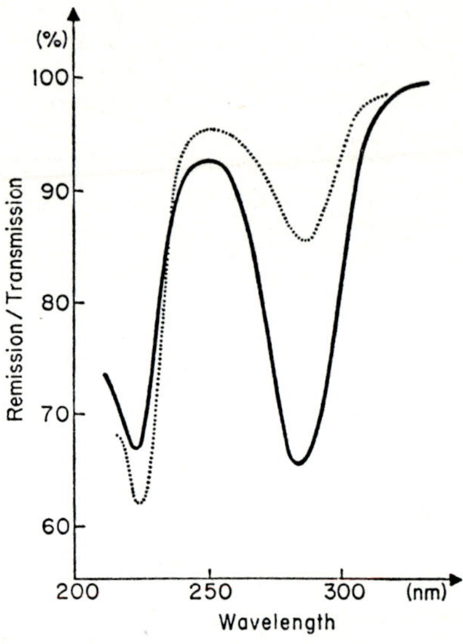

FIG. 5. Absorption spectrum of arbutin. (———) Reflectance (remission) spectrum of
1·3 μg measured directly from TLC plate. (..............) Transmittance (transmission) spectrum
of 28 μg contained in 10 ml. methanol in a cuvette.

For reflectance spectra only 0·2–5 μg of substance are required—spectra
obtained from solution in curvettes require at least ten times as much
substance—so that it is possible to make use of direct determination where
the quantity of substance available is too small for the usual measurements.
Direct evaluation of thin layer chromatograms are thus of great value in the
fields of biological-medical analysis and in criminology.

The question arose, however, as to how a substance which does not absorb
ultraviolet or visible radiation and which is not fluorescent could be observed
and how a direct evaluation of these substances could be made. There are
three possibilities:

(1) Radioactive labelled substance are used and the β rays are determined
directly with appropriate scanning systems (Berthold and Wenzel, 1967).

(2) An indirect measurement of the isotopes can be made by means of
the evaluation of the fluorescence. For this purpose the TLC plates are

sprayed with a suitable scintillation "cocktail" the emission radiation of which can be measured by a photomultiplier (Roucayrol *et al.*, 1963, 1964).

(3) The substances are made to react chemically with absorbent or fluorescent substances. As an example of this method the evaluation of amino acids and peptides can be given. Both groups of substances react with 2·4-dinitro-fluorobenzene, the stoichiometric proportions of which are known. The DNP-derivatives of leucine and arginine, for example, are bright yellow coloured substances, the maxima of both spectra being near to $\lambda = 340$ nm (Fig. 6) with the two curves being almost identical.

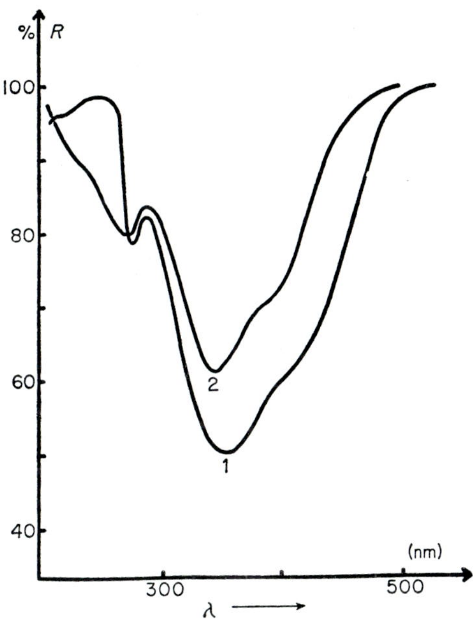

Fig. 6. Reflectance spectrum of DNP-leucine and DNP-arginine measured directly from TLC plate. 1—DNP-leucine, 2—DNP-arginine (1·7 μg in each case).

This is explained by the fact that the absorption is to a large extent determined by the aromatic systems so that a definite characterization of these substances is only possible in association with other criteria, e.g. their R_f values. Similar experiments can be undertaken with phenyl thio-hydantoin derivatives of the amino acids which quench fluorescent light. The practical value of this has been shown by Pataki *et al.* (1966, 1967) and by Baltes (1967) in his work on the DNP-peptides of various kinds of meat. In Germany only beef or veal may be used in the manufacture of packet soups; the addition of mutton, horsemeat or whale meat can easily and rapidly be detected by TLC in combination with the method just described.

G. Quantitative Determinations

It is not only with regard to qualitative methods of evaluation that direct spectrophotometric determination is of interest. Quantitative measurements can be obtained by evaluation with the wavelength of maximum absorption of the substance. This may be illustrated by considering the two chemically similar alkaloids, strychnine and brucine. Neither compound is immediately recognizable on white silica gel layers and if they are rendered visible with the usual alkaloidal spray reagents they would give exactly the same colour. Thus in a densitometric measurement they might be determined together (if not completely separated in the plate) and one might be regarded as the other since they move fairly closely together in most systems. However, by taking a spectrophotometric measurement at a wavelength of 303 nm the brucine can be determined separately from the strychnine and if an additional determination is made at a wavelength of 260 nm the strychnine content can be found by simple subtraction (Jork, 1966).

This method can therefore be used to determine the quantities of all substances separated on TLC plates which absorb ultraviolet light, after constructing suitable calibration curves.

H. Summary

By means of examples taken from pharmaceutical practice it has been shown that direct spectrophotometric evaluation of substances separated on TLC plates possess numerous advantages over other qualitative and quantitative methods of determination. That does not mean to say that the method presents no difficulties and some of these are discussed. The sensitivity of direct absorption measurement is at least ten times greater than that of the processes in which the substance must first be eluted from the adsorbent. The method is also much quicker. Characterization of the separated chromatograms zones by reading the absorption or fluorescence curves directly from the TLC plate was made possible by the introduction of the Chromatogram Spectrophotometer.

REFERENCES

Baltes, W. (1967). Private communication.
Berthold, F. and Wenzel, M. (1967). *Instrumentation in Nuclear Medicine*, **1**, 251.
Bohrmann, H. (1967). Dissertation, Saarbrücken.
Frodyma, M. and Lieu, V. T. (1967). *Anal. Chem.* **39**, 814.
Frodyma, M., Frei, R. W. and Williams, D. J. (1964). *J. Chromatogr.* **13**, 61.
Gänshirt, H. (1967). *In* E. Stahl: "Dünnschicht-Chromatographie, ein Laboratoriumshandbuch", 2nd ed. Springer.
Godin, P. J., King, T. A., Stahl, E. and Pfeifle, J. (1967). *Nature, Lond.* **214**, 319.
Gorbach, G. and Haack, A. (1956). *Microchim. acta* 1751.
Jork, H. (1964). *III. Intern. Symp. f. Chromatogr.*, Brüssel. Symp. Bericht S. 295.
Jork, H. (1966). *Z. analyt. Chem.* **221**, 17.

Jork, H. (1966). *IV. Intern. Symp. f. Chromatogr.*, Brüssel. Symp.-Bericht, S.

Jork, H. (1967). *Cosmo Pharma* 3/1, 33.

Jork, H. (1968). *Pharma International* 4/1, 12.

Jork, H. (1968). *J. Chromatogr.* **33**, 297.

Klaus, R. (1964). *J. Chromatogr.* **16**, 311.

Kortüm, G., Braun, W. and Herzog, G. (1963). *Angew. Chem.* **75**, 653.

Kortüm, G. and Schlichenmaier, V. (1966). *Z. phys. Chem.*, N.F. **48**, 267.

Pataki, G. and Kunz, A. (1966). *J. Chromatogr.* **23**, 465.

Pataki, G. and Niederwieser, A. (1967). *J. Chromatogr.* **29**, 133.

Pfeifle, J. (1967). Dissertation, Saarbrücken.

Roucayrol, J. C. and Taillandier, P. (1963). *C. r. Seanc. Soc. Biol.* **256**, 4653.

Roucayrol, J. C., Bergner, J. A., Meymel, G. and Perrin, J. (1964). *Intern. J. appl. Radiation Isotopes* **15**, 671.

Sawicki, E., Stanley, T. W. and Johnson, H. (1964a). *Microchem. J.* **8**, 257.

Sawicki, E., Stanley, T. W., Elbert, W. C. and Pfaff, J. D. (1964b). *Anal. Chem.* **36**, 497.

Sawicki, E. and Stanley, T. W. (1965). *Anal. Chem.* **37**, 938.

Seiler, N., Werner, G. and Wiechmann, M. (1963). *Naturwissenschaften* **50**, 643.

Stahl, E. (1964). *Arch. Pharmaz.* **297**, 500.

Stahl, E. (1964). *Proc. Soc. anal. Chem.* **1**, 121.

Stahl, E. and Jork, H. (1966). *Arch. Pharmaz.* **299**, 670.

Stahl, E. and Jork, H. (1968). Zeiss Informationen No. 68.

Struck, H., Karg, H. and Jork, H. (1968). *J. Chromatogr.* (in press).

Zeiss, C. (1967). *Druckschrift* No. 50-657/K-d.

Chapter 7

The Application of Spectroscopy to Thin-Layer Chromatography

G. W. GOODMAN

*Research Laboratories, B.P. Chemicals (U.K.) Ltd.,
Epsom, Surrey, England*

Thin-layer chromatography has shown by its rapid expansion over the last decade that it is a technique unequalled for the separation of mixtures of organic compounds from microgram to milligram quantities. To help in the identification of the separated products many ingenious adaptations of organic chemical reactions have been carried out. Such reactions, logical though they may be, have the disadvantage that the zone under test is no longer available in its original form: on the other hand, however, spectroscopic examination of the components of the mixture has the advantage that chemical reactions may be carried out later if desired.

Before the examination of the various physical methods to augment the information obtained from the separated thin-layer zones, there are a number of points concerning the thin-layer separation itself which must be considered. Figure 1 shows the separation of a number of rubber additives on silica gel G with benzene as solvent, the zones being made visible with iodine vapour. The loadings of the various compounds listed have been increased from left to right across the plate. The lowest zone, rosin acid shows an increasing tendency to streak and swamp the adjoining zones as the concentration increases. The streak extends to a degree beyond the area visualized and any solvent system that gives rise to streaking must be immediately suspect in that it may contaminate adjoining zones.

Two further points arise from this plate, the first concerns the amount of material actually present. Visualizing agents such as ultraviolet light or iodine can be very sensitive and may well show up as strong positive zones components present at a concentration of only a microgram or two. The second point is that it is often necessary to alter the solvent considerably in order to remove the more polar materials that remain near the origin on the plate.

The available methods for spectroscopic examination have a great bearing on the amounts of the components of a mixture that it is necessary to separate. A mathematical examination of the available parameters will show that infrared spectroscopy may well require a theoretical sample in excess of 200 μg. If compounds of a similar chemical nature are to be dis-

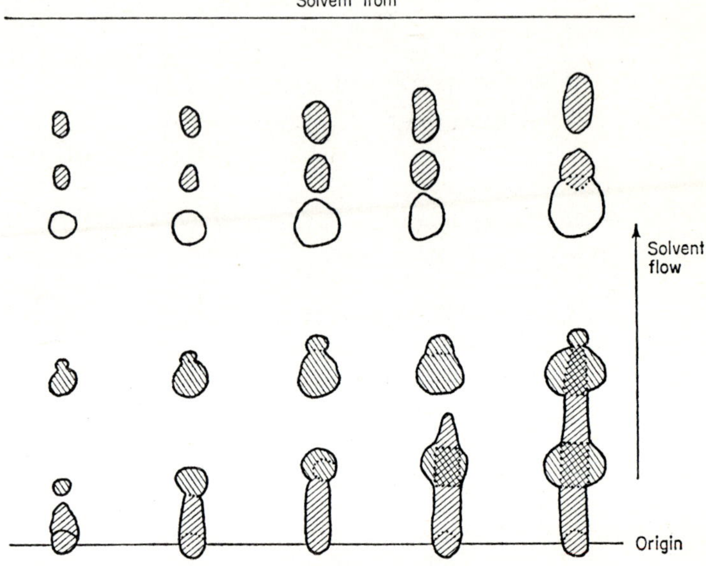

FIG. 1. Effect of concentration on zone area of some rubber additives. ⬤, Rosin acid; ⬤, tetra-ethyl-thiuram disulphide; ○, phenothiazine.

tinguished then peaks with a molecular extinction coefficient of unity must be detectable. This makes the absorbance detection limit about 0·005, assuming that the spectrometer can distinguish between 99% and 100% transmission. Beer's Law gives us that

$$A = \varepsilon \times \text{concentration} \times \text{path length}$$

so that the

$$\text{concentration} = \frac{0 \cdot 005}{\text{length}} \text{ g-moles/l.}$$

or

$$\frac{5}{\text{path length}} \mu\text{moles/ml.}$$

As the length is in centimetres this becomes

$$\text{concentration} = 5 \ \mu\text{moles/cm}^2.$$

Under normal conditions, the area at the focus of the usual spectrometer is about 0·2 cm^2, so that the minimum sample needed at the focus is $5 \times 0 \cdot 2 = 1 \ \mu\text{mole}$.

For a compound of about 200 molecular weight and a density of 1 g/ml. this corresponds to a sample of about 200 μg. If higher values are accepted for the minimum molecular extinction coefficient, which is possible if the substances were different chemically, then the value might be reduced by 10 or even 100. The beam area, or the minimum value of the absorbance also can be reduced. This has enabled infrared spectra to be recorded with 10 μg, or even 1 μg of sample, but this requires considerable expertise and possibly a large measure of luck.

It has been found preferable in this work to have a sample of 100–200 μg where possible, so that the final spectrum is much stronger than that arising from contamination of the sample with silica.

A similar calculation can be applied in the case of ultraviolet spectroscopy, where if it is assumed that the minimum detectable value of the absorbance is 0·04 and the molecular extinction coefficient is about 100, then

$$0\cdot04 = 100 \times \text{concentration} \times \text{path length}$$

which gives, finally,

$$\text{concentration} = 0\cdot4 \ \mu\text{moles/ml.}$$

so that for a cell capacity of 3 ml. the required amount is 1·2 μmoles.

Thus a sample of molecular weight 200 would require about 240 μg. The theoretical amount could, of course be amended by reducing the cell size or scale expansion. Ultraviolet spectroscopy is potentially more sensitive than infrared spectroscopy since the scope for alterations is larger and the molecular extinction coefficients can vary greatly.

Mass spectroscopy will readily handle 1–10 μg although a larger sample is easier to work with. Nuclear magnetic resonance require about 5 mg, but this can be reduced with signal-averaging on a 16-hr run to 100 μg.

Consideration will first be given to infrared spectroscopy within the limits outlined above. Since the amounts on a thin-layer plate are normally small and any technique of solvent extraction could lead to a loss of end product, it was thought desirable to investigate the direct transfer of zones to the potassium bromide.

Before any plates were made, various varieties of potassium bromide were examined and only Spectroscopic grade was found to be suitable. Many ordinary batches appeared to contain ammonia which could not be removed by overnight heating to 250°C and these were rejected; it is possible that they could have been purified by fusion, but it was not felt that the regrinding would have been satisfactory.

A suitable potassium bromide layer was made by grinding the potassium bromide in a glass mortar with ethanol intermittently over a period of about twenty minutes to form a slurry. The slurry in the form of a thin cream was then poured into a small mechanical spreader and applied to the plates which were then left exposed to the air to dry. The potassium bromide formed a hard coherent layer, presumably because the small amount of potassium bromide that had dissolved in the ethanol recrystallized and formed an effective bonding agent between the larger particles and the glass plate surface. The plate was then dried at 105°C for an hour in order to remove traces of alcohol.

Initially, attempts were made to form a simultaneous twin layer plate with a divided spreader applying parallel layers of silica gel and potassium bromide, but difficulty was encountered in forming a good lateral junction and the process was abandoned. The final separation was carried out on a silica gel plate (10 × 20 cm) loading the sample as a series of zones across the plate. When the plate had been run, the right-hand side of the plate was covered with a clean plate, and the left-hand side visualized with iodine vapour. Lines were then scored across the plate with a chisel-edged dissecting needle to mark the boundaries of the unreacted areas and the iodine treated zone scraped off entirely. The potassium bromide plate was then similarly marked into a series of triangles such that their bases coincided with the sample areas on the silica gel plate. The process is illustrated in Fig. 2.

The two plates were then clamped face to face using the device shown in Fig. 3. It was found after a series of runs to be advisable to use as eluting solvent one of the lowest acceptable polarity. This was because traces of silica appeared to travel with the more polar solvents and contaminated the potassium bromide and also because the less polar solvents obviate the risk

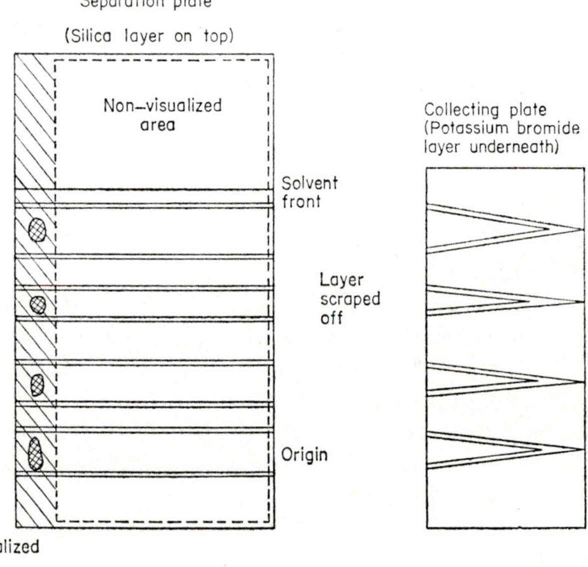

FIG. 2. Plate transfer sequence.

of dissolving the potassium bromide layer. To be on the safe side, it is better
to use solvents that are transparent in the infrared in the region below 7·4
and in the region 11–15. The transfer procedure with the chosen solvent
must be repeated several times to be sure that the zones have in fact arrived
at the points of the potassium bromide triangles. Very often, at this stage,
they can be seen with the naked eye. A check on the adequacy of transfer
can be carried out by finally exposing the silica plate to iodine vapour—it
should of course reveal no traces of the zones.

FIG. 3. Stainless steel plate clamp.

The potassium bromide at this stage holds a moderate amount of eluting
solvent which must be removed before proceeding. Care must be taken to
avoid the loss of sample with the solvent; Fig. 4 shows the surprisingly high
volatility of certain antioxidants when a small sample with a large surface
area is exposed to heat. Satisfactory conditions appear to be exposure to a
pressure of 0·1 mm mercury over phosphorus pentoxide for an hour or two.

When dried the potassium bromide can be removed and pressed into a
disc appropriate to the amount of material available. Sufficient potassium
bromide (50 mg) to fill a 1·0 × 0·2 cm disc former is generally provided
by an area of about 2·0 square cm from a 0·5 mm thick potassium bromide
layer. A possible advantage in using discs with this rather large size slit is
that after examination by infrared spectroscopy the fused potassium bromide
can be readily removed and extracted with solvent for mass spectroscopy.

The interpretation of the final spectra generally is completed by a combination of two methods; the characteristic group frequencies of the molecules are first considered, together with any other information that may be available. This enables a picture of possible molecules present to be built up. Finally the spectra is compared with those of the suggested compounds. The procedure is relatively simple where the range of compounds likely to be present is limited and in these circumstances a very much higher value for the molecular extinction coefficient is acceptable. This, of course, means that a much lower amount of sample can be used. When the nature of the sample is completely unknown, it is often necessary to recover the material from the potassium bromide for a mass spectrometric examination.

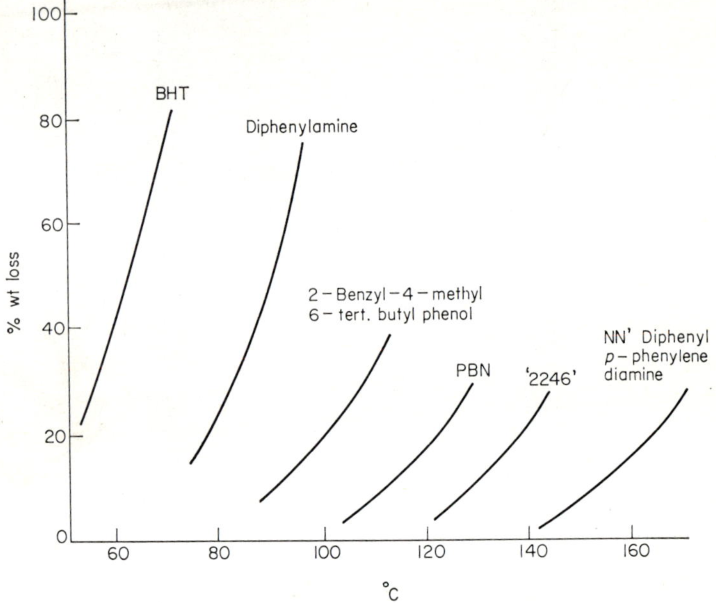

Fig. 4. Relative volatilities of antioxidants.

It was found during a series of trial runs that a certain amount of permanent adsorption occurred on the silica gel plate in some cases. Thus, if a series of loadings of phenyl-β-naphthylamine were made along the 20 cm side of a 10×20 cm plate which was then mounted in contact with a potassium bromide plate as illustrated in Fig. 5, amounts of less than 50 μg were not detected by the method outlined above. The extent to which this effect varies has not been fully investigated and it may well be quite considerable for some more polar molecules.

Ultraviolet spectroscopy is a rather less applicable tool for the identification of organic compounds since only some arrangements of atoms give rise to a useful absorption. Nevertheless, the electronic structures of a number of types of compounds can give valuable leads on the natures of the materials present. Amongst these compounds are aldehydes, ketones, aromatic compounds and also surfactants which will be dealt with in more detail later.

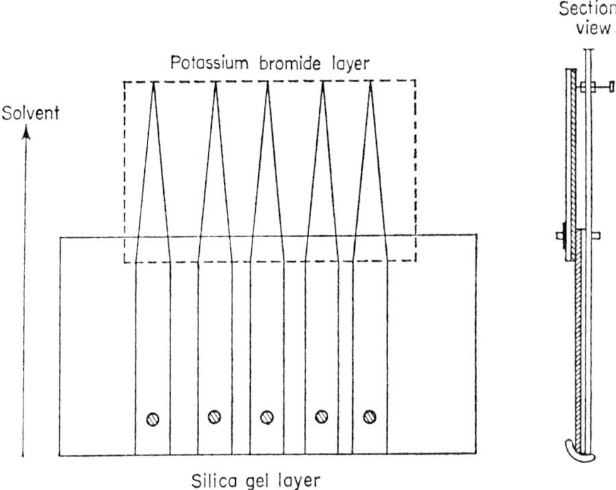

FIG. 5. Retention of PBN on silica gel G.

Reid *et al.* (1955) published an excellent article in which they identified a large range of surface active agents by means of their ultraviolet spectra. Implicit in their identification was the use of a pure sample, for the preparation of which thin-layer chromatography is ideally suited. It must be stressed here that, unlike working with infrared spectroscopy, it is essential to have a clear idea of the types of compounds between which it is desired to differentiate.

The zones produced by surfactants lie near the origin on adsorption plates run with a non-polar solvent and clearly cannot easily be transferred to a potassium bromide disc by the methods outlined above. If, however, the zone can be separated from its neighbours then ultraviolet spectroscopy can be used. Theory predicts that using a 3 ml. cell, less than about 100 μg should produce a reasonable spectrum of the benzenoid bands of an aromatic or heteroaromatic compound. This, however, is unfortunately not the case, since these more polar compounds are very strongly adsorbed on the silica gel. In order to obtain sufficient material for a worthwhile spectrum it is necessary to run either a preparative loading, or if the other substances in the mixture have suitable R_f values, perhaps an overload and "run off" plate.

It was found that in order to differentiate between rosin acid and sodium oleate, it is necessary to load about 5 mg on the plate. The effect of the silica is very clearly seen in Fig. 6 in the case of sodium dodecyl benzene sulphonate. In the case of the upper curve 5 mg of the sodium salt were scraped off the plate in about 1 g of silica gel, shaken with water in a centrifuge tube, then centrifuged and the clear supernatant liquid examined in a Unicam SP800 using 1 cm silica cells. The lower curve was obtained by identical treatment except that in this case there were 2 g of silica present. The absorption bands of the sodium dodecyl benzene sulphonate can readily be distinguished with maxima at around 2250–2300 and the fine structure at around 2500–2800 Å. Sodium oleate, on the other hand, has a maximum at 2250 only.

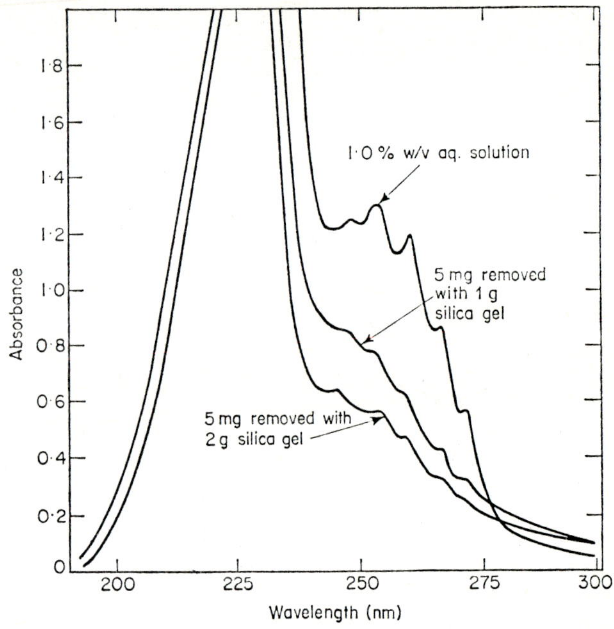

FIG. 6. Adsorption of sodium dodecyl benzene sulphonate on silica gel G.

Although these examples have been quoted using an aqueous extract, quite a wide range of solvents appear to be satisfactory as seen in Fig. 7.

Trespassing slightly into the realm of chemical reactions, it is interesting to note that phenols can be additionally confirmed by examining the spectra of neutral and alkaline solutions. If this is done, a bathochromic shift is observed in λ_{max} and an increase in ε_{max}. This change arises because an additional pair of non-bonding electrons are available for interaction with the π electrons of the aromatic nucleus. Similar confirmatory tests can be carried out with aniline derivatives in neutral and acid solutions.

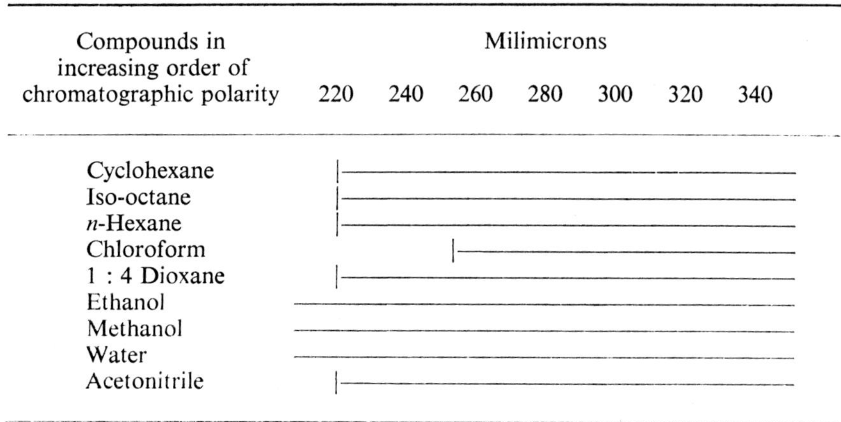

FIG. 7. Polarity/transparency ranges in the near ultraviolet.

The other instrument that is of considerable use on thin layer plate extracts is the mass spectrometer because of the small weight of sample that it will accept. Since this is a "destructive" technique, in that the sample cannot be recovered, it has been left until last. It is a natural "follow-on" to infrared spectroscopy where it has proved useful in the elucidation of mixtures of hindered phenols.

It is, of course, appreciated that the use of these techniques is rather akin to using a sledge hammer to crack a nut, but they *can* provide unequivocal evidence on the components of a mixture without having recourse to a scaling-up process to obtain gram quantities. With the increasing availability of spectrometric equipment, considerable value can be obtained from them.

REFERENCE

Reid, V. W., Alston, T. and Young, B. W. (1955). *Analyst* **80**, 682

Chapter 8

The Visual Assessment of Thin-Layer Chromatograms

C. A. JOHNSON

*Scientific Director, British Pharmacopœia Commission,
London, England*

Any survey of methods for the quantitative use of thin layer chromatography must include some reference to the first-used and simplest procedure of all; that is, the system in which the only detector is the human eye, the only computer the human brain. Because it is so simple and has been so widely used it may seem that there is little to say on the subject and that what can be said is already well known. This may be so, yet it would seem worthwhile to review some of the difficulties and pitfalls of the method, especially as a number of the points to be made apply equally well to some of the more mechanized techniques that are available. Since interest often lies in the estimation of small amounts of impurities in organic compounds it is from that standpoint that the subject will be reviewed.

There are two purposes for which the method may be used:

(a) to ensure that impurities present are below a certain value, that is, as a limit test, and

(b) to estimate the actual quantity of minor ingredient present, that is, as a semiquantitative assay procedure.

To achieve the first of these aims a very commonly used procedure is to apply a definite amount of substance in a prescribed volume of a specified solvent. After chromatography and revelation of the chromatogram there should be only one spot showing. This is undoubtedly the very simplest approach that could be made to the semiquantitative use of thin-layer chromatography, but it has many shortcomings. For this type of criterion it is necessary to work out a loading that will yield only one spot if the substance being examined is just sufficiently pure to be acceptable, but that will begin to show secondary spots as soon as the allowable limit is exceeded. The principal shortcomings of this approach are as follows.

(a) The technique of spotting must be uniform from laboratory to laboratory. Although this may be achieved by a single worker, or even by a group of workers in a single laboratory, it is impossible to lay down conditions that will guarantee interlaboratory uniformity. A given volume of solvent can be applied with many different devices and varied techniques and it is quite possible to obtain spot diameters from about 4 mm to 10 mm and thus spot areas that differ by a factor of about 6. Obviously, when looking for a spot at the limit of its sensitivity the area over which it is spread is of great significance.

(b) Depending on the type of revealing agent used, the efficiency with which it is applied to the plate will influence the sensitivity achieved.

(c) According to the distance that the impurity has moved, so will its susceptibility to detection vary. An impurity that moves only a short distance from the spotting line will diffuse much less than one which moves almost with the solvent front and the former will thus be revealed at a lower level than the latter. Assuming equal susceptibility to the revealing agent the difference in detectability, in an extreme case, might be about threefold for a 10 cm run.

(d) The criterion that a secondary spot can or cannot be seen is an extremely subjective one; an analyst who knows where a secondary spot *might* appear may well see a spot that would not be observed by a less informed worker. There is, too, a natural tendency to spray a little more thoroughly on that area of a plate where a secondary spot might be expected. Or, if the operator would rather see the sample accepted than rejected, to spray a little less thoroughly. This is a purely psychological effect, but it has real significance.

On the whole this type of all-or-none criterion is a most unsatisfactory one and should not be used.

A second approach that has been used is to apply the sample at two levels, the lower being that at which it is intended to limit impurities. For example, with ergometrine maleate it is possible to apply a loading of about 20 μg and a second loading on the same plate of 0·2 μg. When the chromatograms are revealed by spraying with acid dimethylaminobenzaldehyde any subsidiary spots on the 20 μg chromatogram can be compared with the 0·2 μg spot of ergometrine to determine whether they are greater or smaller. If greater the assumption may be made that the subsidiary spot is present in excess of 1% of the main constituent. This approach suffers from two main deficiencies.

(a) An impurity spot will have moved a different distance on the plate than the control spot and thus will have suffered a greater or lesser degree of diffusion.

(b) The method is only valid (and then only very approximately) if it is known that likely impurities will have a similar sensitivity to the revealing agent as the "parent" substance. With the ergot alkaloids this is a reasonable assumption to make but with many other classes of substances and other revealing agents the approach is quite impossible. For example, oestrogenic steroids are readily detectable by spraying the plate with 10% sulphuric acid in ethanol, heating for about 15 min. at 130° and then examining with long-wave ultraviolet light. One per cent of oestrone in ethinyloestradiol would show as a pale blue, weakly-fluorescing spot; one per cent ethinyloestradiol in oestrone, however, would show as a bright red intensely-fluorescing spot that might lead one to suppose it was the major constituent of the mixture. This criticism is, of course, equally applicable when results are obtained with scanning devices by comparing areas of peaks derived from different components on the same chromatogram.

A far more satisfactory method than the two already discussed is to apply a standard quantity of the known impurity which is sought and to compare it with any corresponding spot in the sample under examination. This is a more difficult thing to do, since it demands a knowledge of the identity of the impurity and a supply of that substance for use as a comparison standard. Ideally, too, the spot of the impurity should be added to a spot of a highly purified sample of the material being examined, applied to the plate at the same loading, since the presence of a large excess of the main substance can sometimes modify the chromatographic behaviour of the small amount of impurity. Usually this modification is confined to a change in the resulting spot area or shape, but in extreme cases it may also affect the distance that the impurity runs. This is the case, for example, with dichlorophen, in which 4-chlorophenol is a probable impurity. In the presence of a 99-fold excess of dichlorophen the chlorophenol moves in, for example, toluene, with an R_f of about 0·2 on a silica gel plate. In the absence of the main constituent the same quantity of chlorophenol will have an R_f of only about 0·1.

For semiquantitative estimation of the level of an impurity it is necessary to apply a series of standards. To achieve the most reliable results it is clearly desirable to design the system of chromatography so that, if possible, the required spot is round and uniform and is well-separated from other spots, particularly the one due to the main constituent. Provided the spots applied to a plate are uniform in size, the question of absolute spot size is of less importance than with the "all-or-none" test. Indeed, a somewhat larger area of spot, for a given amount of substance, might be desirable since the more intense, compact spot might produce errors due to varying concentration in depth on the plate.

If a choice of spray reagents is available for a given determination it is well worth while, in designing the method, to carry out comparative tests with the alternatives. Clearly, the greater the sensitivity of a reagent, the greater will be the difference in colour for standard increments of substance and so the more precisely can an estimation be made. Where reagents are of approximately equal sensitivity but give different colours the optimum choice may be a purely personal one. It is more satisfactory to compare pure colours, such as red or blue, than mixed colours, such as violet or orange. The comparison of browns such as those produced by charring techniques, is, from the point of view of accurate gradation, the most difficult of all.

Another factor that must influence the choice of revealing agent is the complexity of the chemical reaction necessary to produce the required result. It is not a question of time alone that will favour the choice of a one-stage procedure, for each additional stage may introduce its own possibilities of error.

Visualization by a colour-producing reagent is not, of course, the only method available to the analyst. Certain spray techniques result in spots that fluoresce when they are examined with a long wave ultraviolet source. In this case it is the intensity of fluorescence that must be assessed and similar considerations apply to those discussed for coloured spots. Fluorescence methods of this type frequently have the advantage of a higher degree of sensitivity than is associated with colour reactions but it is not everyone who has the ability to grade intensity of fluorescence with sufficient accuracy. A very popular method of visualization is to use an adsorbent, such as the so-called GF 254, impregnated with a fluorescent compound. When examined under short-wave ultraviolet light spots are seen as dark are as on a strongly fluorescent background. This technique has the considerable theoretical advantage that no spray reaction is involved and thus a considerable source of possible error is eliminated. However, the shortcomings of the method far outweight this advantage. Quite apart from the facts that the grading of dark spots is difficult and that the sensitivity is not great, a considerable drawback is that, having studied the plate for a second or two, the worker may after flicking his eyes or blinking see two sets of spots when he looks back to the plate. This effect is multiplied with further examination and a whole series of spots, each of different intensity as the images fade, can be seen. This is the "measles" effect. The plate with fluorescent background has a significant part to play in thin-layer chromatography, but not for the quantitative assessment of low levels of impurities.

Having decided on the most suitable means of visualization the next important step is to decide the range of concentrations that the applications of standard should cover. In the first case this will depend on the sensitivity of the reagent; the more sensitive the reagent the lower will the range need to be and the closer can be the increments between successive spots. In general the lowest concentration used as a standard should be just visible and the highest should contain an amount of standard that is about four or five times that of the lowest. The number of increments used to achieve this range might be about eight or ten; more increments than this are unlikely to produce successive increases in spot intensity of a sufficient amount to allow them to be unerringly differentiated on all occasions. The optimum amount of sample to be applied, which may have to be decided on the basis of a preliminary run, should yield an amount of the constituent to be determined that falls within the range; preferably the sample application should be adjusted to produce an impurity spot that falls within the lower part of the range, for it is in this region that the greatest difference exists between successive increments. For example, in a series of standards at 0·1 μg intervals between 0·1 and 1·0 μg the difference between spots 1 and 2 will be 100%, whereas that between spots 9 and 10 will be only about 10%. In addition to this point it seems to be true that, for the majority of observers, it is easier to assess differences between weaker colours than between those that are more intense.

These generalizations are exemplified by some collaborative work carried out on the determination of 4-chloroacetanilide in phenacetin. The details of the procedure used are not of importance; the optimum range for standards derived from 4-chloroacetanilide was found to be from 0·2 to 0·8 μg at 0·1 μg intervals. The amount of sample taken was such that the 0·2 μg standard was equivalent to 0·01% of 4-chloroacetanilide in the phenacetin being examined. The results in Table 1 are typical of the kind of agreement achieved. Better agreement was evident where the comparison spot was in the lower part of the recommended range. If the amount of sample applied for, say, sample 5, is reduced two or threefold the agreement is improved.

TABLE 1

Percentage of 4-chloroacetanilide found in samples of Phenacetin

Sample no.	Laboratory				
	A	B	C	D	E
1	<0/01	<0·01	<0·01	0·005	—
2	0·01	—	0·01	—	0·02
3	0·02 0·03	0·04 0·03	0·02	0·03	—
4	0·07	0·05	0·07	0·07	0·06
5	0·04 0·05	0·08	0·07	0·06	—

The spotting plan may also be designed to promote easier assessment. It is, for example, unwise to apply the unknown sample as spot number 1 followed by a series of standards in ascending order. Perhaps the soundest scheme is to apply increasing quantities of standard on alternate starting positions on the plate and to apply the chosen amount of sample solution on each of the intermediate positions. This enables a direct comparison to be made between unknown and standard spots side by side without the need to move and refocus the eyes. A simple masking device may be fashioned to cover the whole of the plate except the area in which the spots being compared are located, thus enabling the comparison to be made without peripheral distraction.

In making assessments it is worthwhile remembering that the final comparison may be made by someone other than the worker who has carried out the chromatography. Even better, the comparisons may be made, without collusion, by several observers. This helps to reduce the tendency to human bias that is inevitable with so subjective a method. It is, too, a well-known, or at least oft-repeated, fact that the feminine observer has a greater appreciation of colour shade and intensity than the male; so at this stage of the determination the laboratory secretary may give a better assessment of the plate than the chief analyst. In parenthesis here it might be appropriate to make a plea that all male analysts be subjected to a test for colour-blindness; this is routine in many organizations, but overlooked in others. The colour-blind analyst can carry out many duties in the laboratory but the visual examination of thin-layer chromatograms should not be one of them.

The author submits that with a properly designed test and with care and attention in applying it, the visual assessment of thin-layer chromatograms should always be the first method to be tried. In assessing trace impurities it matters little if an error of 20% is incurred. And the mere fact that a given piece of apparatus might print out a result to two places of decimals is no guarantee that the figures have any validity. It has been said before in other connections, and is equally true here, that it is better to be approximately right than precisely wrong.

Chapter 9

Quantitative Scanning of Radioactive Thin-Layer Chromatograms

B. A. WOOD

*Department of Drug Metabolism, Pfizer Limited,
Sandwich, Kent, England*

INTRODUCTION

Thin-layer chromatography (TLC) has several advantages over other forms of liquid-solid chromatography, the two most important being speed and resolving power. These features should also apply to the method used to detect compounds separated by TLC. When the separated compounds are radioactive, three main methods of detection are available: (a) autoradiography; (b) liquid scintillation counting; (c) chromatogram scanning.

Autoradiography, i.e. exposing X-ray film to the developed chromatogram, provides resolution comparable to that of the original chromatogram, but exposure times can vary from several hours to several days or weeks, depending on the level of radioactivity. Quantitative analysis is achieved by scanning the developed film with a densitometer.

Liquid scintillation counting requires sections of the adsorbent to be transferred from the chromatogram into vials, addition of scintillator, and counting. The counting process is efficient, and therefore fast, giving quantitative results. The resolution is limited by the width of the sections taken for counting, and unless a large number of samples are taken, resolution will be poor. Transferring the adsorbent from the plate to the vials quantitatively is slow and tedious, although an apparatus for performing this task automatically has been described (Snyder, 1964, 1965).

The third method of detection uses a radiation sensitive device moving relative to the chromatogram, producing a signal which is recorded to give a peak or peaks corresponding to the separated compounds. Resolution can be varied by changing the detector slit width; the time required to scan a chromatogram is comparatively short; and quantitative results can be obtained from measurements of peak area (Williams and Smith, 1951; Ravenhill and James, 1967). In addition, the chromatogram is kept intact for subsequent examination by chemical spray reagents.

This chapter discusses the results of an examination of factors which must be considered when quantitative results are required from scanning a radioactive thin-layer chromatogram. The instrument used in this work is the Desaga Thin-Layer Scanner, with a gas flow detector, and has been described elsewhere (Wilde, 1965). The radioactive samples were prepared using [^{14}C]labelled compounds.

Although the results and conclusions apply specifically to thin-layer chromatograms, some of them also apply to paper chromatograms.

A. COUNTING STATISTICS

The disintegration of radioactive materials is Poissonian. If the Poisson distribution is applied to a system recording more than about 100 events, the distribution approximates to a special case of the normal Gaussian distribution, and it can be shown that for a large number of observations the error of the mean value is given by

$$e = k\sqrt{N} \tag{1}$$

where e = error

N = mean number of events recorded

k = a constant.

If $k = 1$, the error is the standard deviation, i.e. 68% of the observations will have an error less than or equal to the computed error. The error in this case is regarded as having a confidence level of 68%. If a higher confidence level is required, then the constant k must be larger. When $k = 2$, the confidence level is about 95%; when $k = 0.67$, the confidence level is 50%.

The error may be expressed as a percentage

$$x = \frac{100k}{\sqrt{N}} \tag{2}$$

where x = percentage error.

As an example of the use of this equation, if the level of radioactivity in a sample is required with an error of 1% at a confidence level of 95%, then 40 000 counts must be recorded.

This account of the statistics of counting is very elementary. The subject has been treated more thoroughly in other references, which also give the variation in confidence level with k (Tittle, 1962; Wang and Willis, 1965).

B. EFFECT OF SCAN SPEED

From equations (1) and (2) it is evident that for accurate results a large number of counts must be collected. This requires a slow scan speed. An approximate indication of the scan speed required, may be obtained if consideration is given to a radioactive zone having a distribution of radioactivity as shown in Fig. 1. If this zone is scanned by a detector having a slit length perpendicular to the direction of travel greater than the length of the zone in that direction, then it can be shown that

$$N = \frac{wDE(w+l)}{100lv} \tag{3}$$

where N = number of counts recorded
    W = effective slit width (cm)
    D = total radioactivity in zone (d.p.m.)
    E = detector efficiency (%)
    l = length of zone in direction of scan (cm)
    v = scan speed (cm/min).

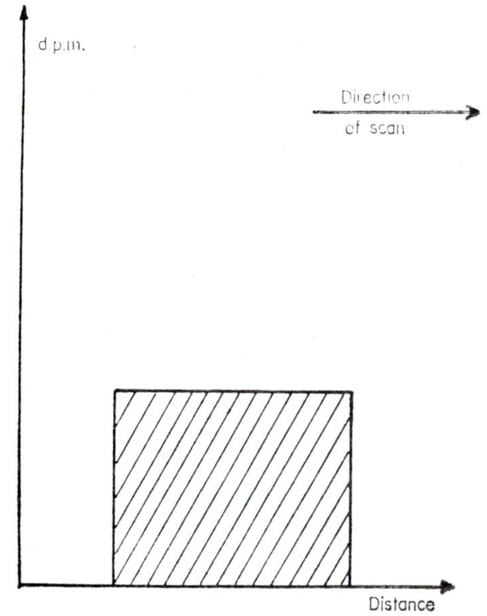

Fig. 1. See page 110.

Thus a high value of N can be obtained by using a wide slit, a high level of radioactivity, a high detector efficiency, or a slow scan speed. In practice, the only parameter which is variable to any significant degree is the scan speed. Rearranging equation (3) and substituting for N from equation (2)

$$V = \frac{wx^2DE(w+l)}{10^6\,lk^2} \quad \text{cm/min.} \tag{4}$$

Therefore, if a 6 mm diameter spot containing 10 000 d.p.m. is scanned by a detector of efficiency 20% having an effective slit width of 4 mm, and the result is required with an error of 1% at the 68% confidence level, then from equation (4) the scan speed must be 8 cm/hr.

Equation (4) was derived assuming a distribution of radioactivity as shown in Fig. 1. In practice the distribution will approximate to a Gaussian curve, as shown in Fig. 2, which is drawn so that the shaded area is the same in Figs 1 and 2. It can be seen from Fig. 2 that although most of the radioactivity is concentrated over a distance less than l, the count rate in this region is higher than that occurring in the hypothetical system of Fig. 1. The increased count rate will tend to compensate for the shorter time available for counting, so that equation (4) probably represents a reasonable approximation.

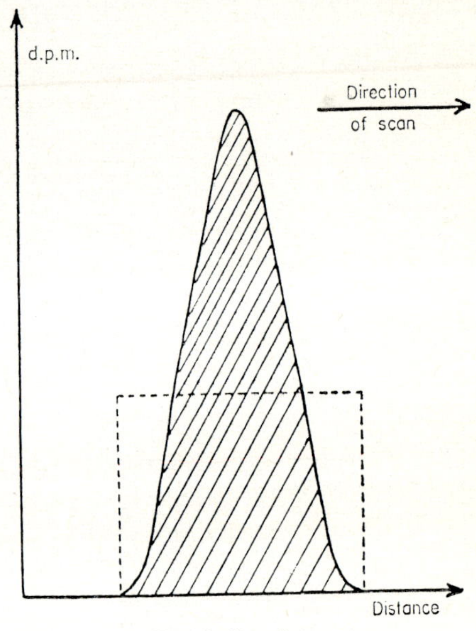

Fig. 2. See above.

It is not intended that equation (4) is applied before scanning every chromatogram. The main purpose is to emphasize the need for slow scan speeds if quantitative results are required.

C. Effect of Detector Voltage

The gas flow detector requires a high voltage to be applied between the anode wire and the cathode. The magnitude of the response from the detector, and hence the detector efficiency, varies with the applied voltage. The optimum voltage can be determined by placing the detector over a radioactive source, and plotting the voltage/response curve. The voltage/response curves for two gas mixtures are shown in Figs 3 and 4. The response increases with voltage until a plateau is reached. As the voltage is further increased, instability occurs due to arcing in the detector. For maximum efficiency and stability, the voltage should be set to that corresponding to the centre of the plateau. The voltage at which instability occurs appears to depend on the level of radioactivity in the sample, being lower for samples of high activity. For the 98·5% argon, 1·5% propane mixture, this voltage has been found to vary by 160 volts (range 1380–1540 V), depending on sample.

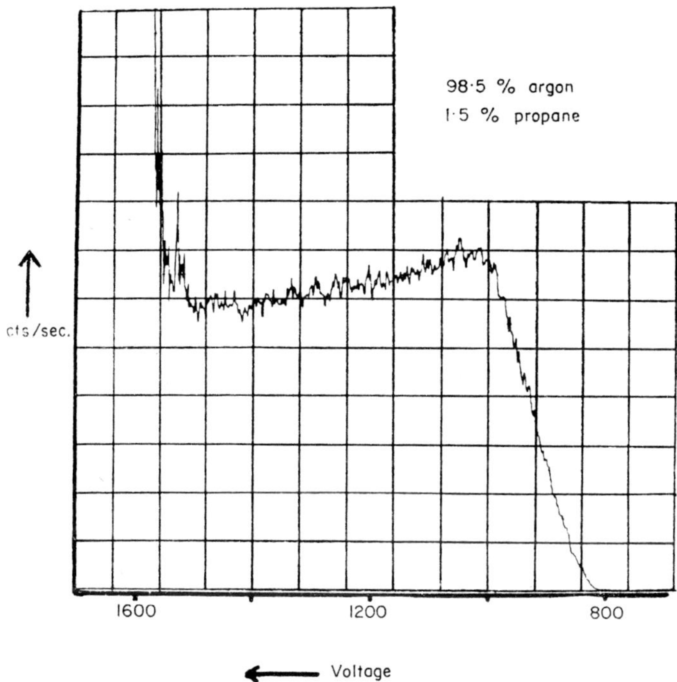

FIG. 3. Voltage/response curve for 90% argon, 10% methane. Chart speed, 30 cm/hr; full scale deflection, 300 cts/sec; time constant, 3 sec. Voltage increased at 40 V/min.

The reason for the slight maximum observed with 98·5% argon, 1·5% propane mixture is not known. It is reproducible, and independent of the sample radioactivity.

There is no significant difference in sensitivity between the two gases, and the argon/propane mixture was used in the following work.

D. Effect of Detector Height

The detector of the Desaga Scanner can be adjusted to vary the height of the detector above the chromatogram and in order to determine the effect of detector height (i.e. the distance between chromatogram and detector) on peak shape, a radioactive sample was scanned at detector heights ranging from 0·25 to 2·00 mm. The sample was prepared by isolating a section of silica gel 5 mm × 2 mm on Eastman Chromagram sheet, and applying radioactive solution to the isolated area. The detector height was set by means of feeler gauges.

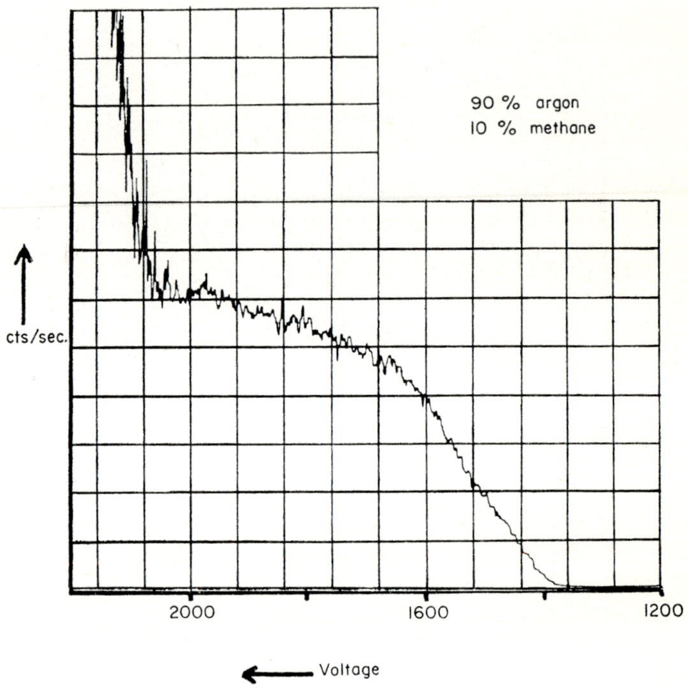

FIG. 4. Voltage/response curve for 98·5% argon, 1·5% propane. Other details as for Fig. 3.

The scans for 0·25 mm and 2 mm spacings are shown in Fig. 5(a) and (b). Figures 6, 7 and 8 show the variation of peak height, peak half band width, and peak area, with detector height. From Figs 6 and 7, it is clear that for maximum peak height, and minimum peak width, the detector must be as close to the chromatogram as possible. However, the closer the detector is to the plate, the greater the risk of contamination with silica from the chromatogram. A reasonable compromise would be 0·5 mm between detector and chromatogram.

From Fig. 8, it can be seen that there is a slight increase in peak area (i.e. number of counts recorded) with detector height. This increase is small compared to the change in peak height and half band width, and is discussed later.

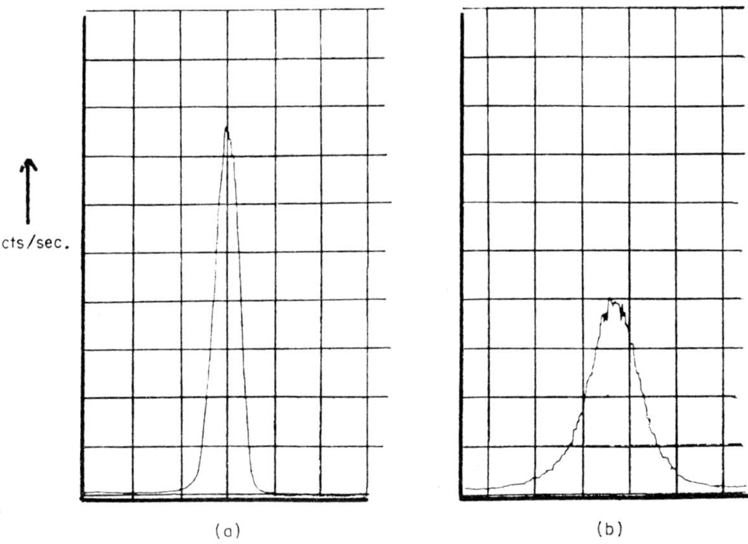

FIG. 5. Effect of detector height on peak height and peak width. The black area above the scans indicates the size of the radioactive sample. Scan speed, 12 cm/hr; chart speed, 30 cm/hr; full scale deflection, 300 cts/sec; time constant, 3 sec. (a) detector height, 0·25 mm., (b) detector height, 2·00 mm.

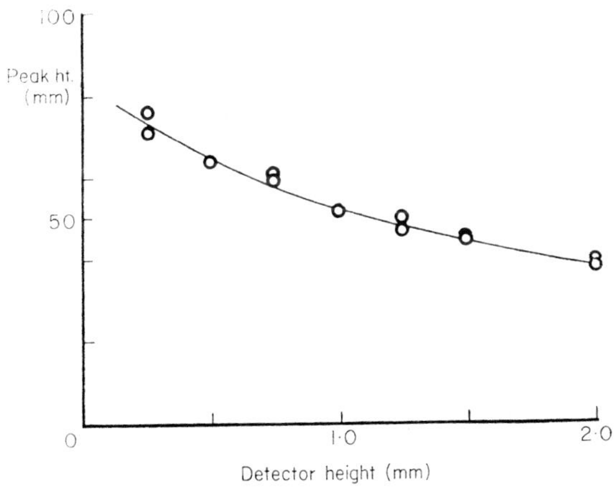

FIG. 6. Variation of peak height with detector height.

    B. A. WOOD

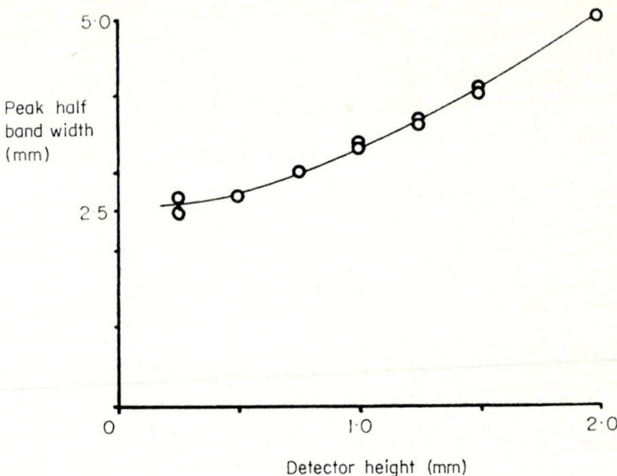

Fig. 7. Variation of peak half band width (width of peak at half the peak height) with detector height.

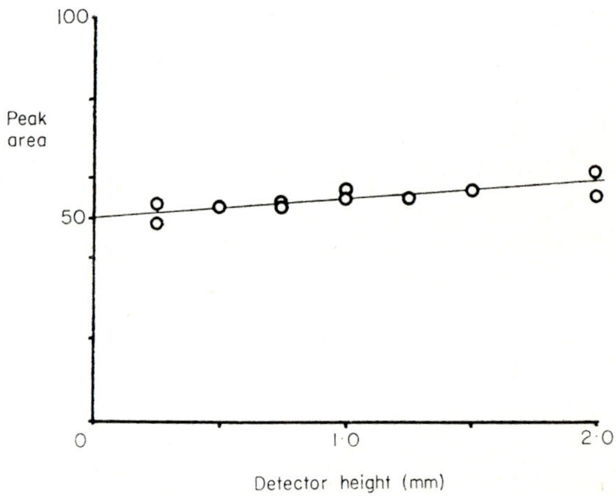

Fig. 8. Variation of peak area with detector height. Equation of line, by method of least squares, is: peak area = 4·4 (dectector ht.) + 50·6. Peak area determined by cutting out and weighing a photocopy of the peak. The ordinate is the weight of the peak, in mg. 50 mg is equivalent to about 5700 counts.

E. DETECTOR EFFICIENCY

The efficiency of a detector is the ratio of the number of counts recorded in unit time to the number of β-particles entering the sensitive volume of the detector in unit time. This depends on the design of the detector, and is constant under given conditions of voltage and counting gas. When scanning chromatograms, the apparent efficiency will be less than the true efficiency if there is any factor which reduces the number of β-particles entering the detector.

From equation (3), the number of counts recorded in unit time is related to the effective slit width, so that to determine the efficiency by scanning a sample of known activity it is necessary to know the effective slit width and assume that equation (3) is valid. If the detector is held stationary over a spot of known activity, and if the slit dimensions are such that the detector "sees" the whole sample, then the slit width does not enter into the calculations.

The efficiency of the Desaga detector was determined by placing the detector, fitted with a 4×10 mm slit, over an area of silica gel 4×8 mm, to which $2\,\mu$l. of a $[^{14}C]$glucose solution (99,200 d.p.m./μl.) had been applied. The signal was recorded for at least 10 minutes. Five samples were measured, and for one sample the variation in efficiency with detector height was determined.

The average efficiency measured in this way was $17 \cdot 4\%$ for a detector height of $0 \cdot 5$ mm. This figure for the efficiency takes into account the factors of backscattering, and absorption by air and silica gel. The silica gel layer was nominally 250 μ thick, and the weight of silica gel per cm^2 was $9 \cdot 9$ mg. Assuming a density of $2 \cdot 2$ g-cm^3 for silica, this is equivalent to a layer of silica about 45 μ thick. The average range in glass of β-particles from ^{14}C is about 100 μ, so that there will be comparatively little loss by absorption in the silica gel.

Increasing the detector height decreased the efficiency by about 1% per $0 \cdot 5$ mm for the range $0 \cdot 5$–$2 \cdot 0$ mm (i.e. at 2 mm detector height, the efficiency was about 14%).

F. EFFECTIVE SLIT WIDTH

In the experiments described under "Effect of Detector Height" the peak area, which is proportional to the efficiency, was found to increase with detector height, while the experiment described under Section *E* showed the efficiency to decrease with detector height.

Radiation from a radioactive sample is emitted in all directions, and some radiation will be emitted in a direction very nearly parallel to the plate. When the detector is stationary, and separated from the plate by a short distance this radiation is not detected. The greater the detector height, the greater the loss, giving rise to a decrease in apparent efficiency. When the detector is scanning, there is a finite probability that radiation emitted nearly parallel to the plate will enter the sensitive volume of the detector when the detector is not over the spot, resulting in an increase in the number of counts recorded. This effect will increase with detector height.

In Fig. 5(b), the radioactive area is 2 mm wide, and the nominal slit width is 2 mm, but the detector starts recording counts about 8 mm away from the centre of the spot. The effect is that of using a larger slit.

Theoretically, the base width of a peak should be $(w+l)$ cm, so that the effective slit width should be calculable from the peak base width. It is difficult to measure the base width for a peak such as that in Fig. 5(b), but a rough estimate can be made, and the results are plotted in Fig. 9. The ordinate represents effective slit width calculated assuming the peak base width is $(w+l)$ cm. At a detector height of 1·5 mm the effective width is 12 mm, although the nominal slit width is only 2 mm.

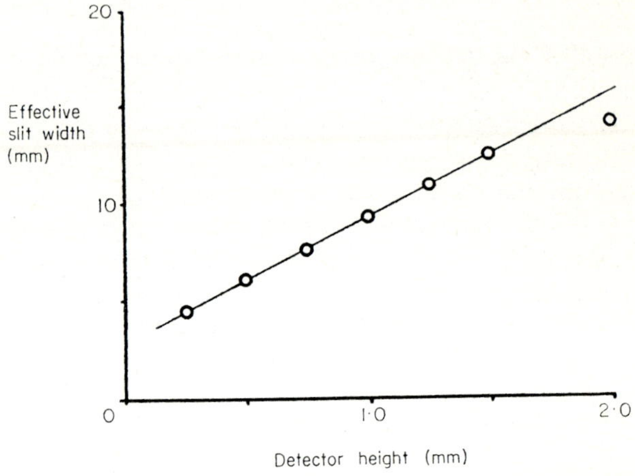

FIG. 9. Variation of effective slit width, calculated from the peak base width, with detector height.

Use of a large slit has the advantage that more counts are collected than when a narrow slit is employed. This advantage is not realized when a narrow slit is used with a large detector height since only a marginal increase in peak area is observed as the detector height is increased (Fig. 8). It is evident that the effective slit width considered in terms of the peak base width is different to the effective slit width considered in terms of collecting counts. In the first case the effective slit width is very dependent on detector height, but in the second case the effective slit width is not very dependent on detector height, and approximates to the nominal slit width.

G. RESOLUTION

If two radioactive zones are separated by a distance d cm, then if $d > w$, they will be completely resolved, and will appear as two separate peaks. If $d < w$, the two peaks will be recorded as partially overlapping. The height of the minimum between the two peaks, expressed as a percentage of the mean height of the two peaks will be called the valley height. When the valley height is 100%, the peaks are not resolved; when the valley height is 0%, the peaks are completely resolved.

A sample consisting of five areas of silica gel 3 × 5 mm, each containing the same amount of radioactivity, and separated by 1, 2, 3 and 4 mm gaps, was scanned using slit widths of 1 mm, 2 mm and 4 mm. A similar technique has been used previously in work with paper chromatograms (Bonet-Maury, 1953). In Fig. 10, the valley height is plotted against peak separation. From this it can be seen that even with a 4 mm slit, peaks separated by 2 mm will be partially resolved, so that for qualitative work, where only the number and position of peaks is required, quite wide slits should be adequate. For quantitative work where complete separation of peaks is required, the smallest available slit and minimum detector height should be used.

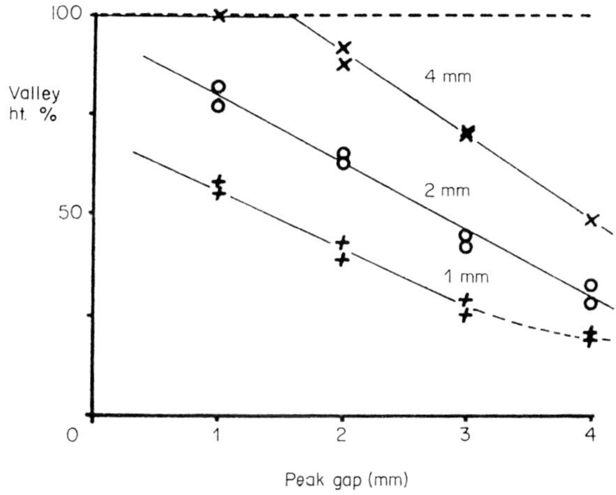

FIG. 10. Variation of valley height (defined in text) with gap between 3 mm × 5 mm areas of radioactivity (peak gap), for 1 mm, 2 mm and 4 mm slits.

It is tempting to extrapolate the plots of Fig. 10 to give the effective slit width (valley height = 0), and the minimum peak separation that can be resolved (valley height = 100%). Unfortunately, there are insufficient points to justify a linear extrapolation, as is evident from the curve for the 1 mm slit.

H. TIME CONSTANT

The main cause of "noise" on the recorder trace is statistical fluctuation in the level of radioactivity of the sample being measured. This noise can be reduced by the use of a time constant in the recorder circuit.

The statistical fluctuation will be proportional to $\sqrt{N}$ (equation (1)), so that the signal to noise ratio will increase as N increases (being proportional to $1/\sqrt{N}$, equation (2)). Thus the time constant can be reduced as the signal level increases.

The use of a time constant produces distortion of the trace: the peak height is reduced, the band becomes wider, and the peak maximum is displaced (Fig. 11). This distortion will increase as the scan speed increases so that to keep distortion to a minimum, the time constant should be as low as possible consistent with an acceptable noise level on the recording.

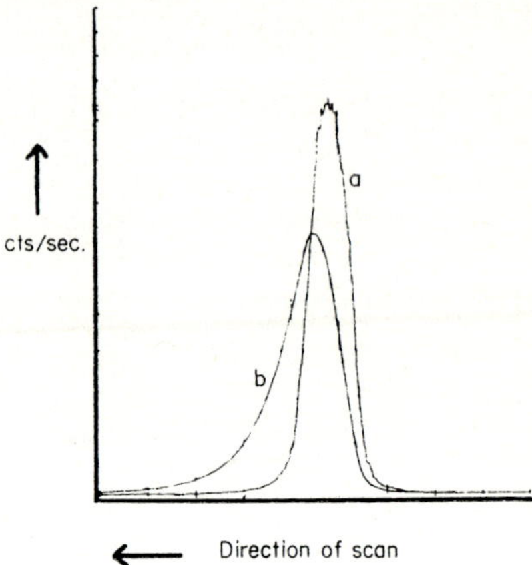

FIG. 11. Effect of time constant. Curve a, 1 second time constant; curve b, 30 second time constant. Scan speeds, 30 cm/hr; chart speed, 60 cm/hr; full scale deflection, 300 cts/sec.

I. CONCLUSION

Factors affecting the scanning of radioactive thin-layer chromatograms for subsequent quantitative analysis have been investigated. For optimum results, it has been shown that the main requirements are (i) a low scan speed, to enable the maximum number of counts to be recorded; (ii) a narrow slit, for maximum peak separation; and (iii) the minimum distance between chromatogram and detector, to improve resolution and detector efficiency.

These requirements are probably well known, but the present work has attempted to provide some experimental evidence for these conclusions. The results obtained suggest that there is much more work to be done before all the factors involved are fully understood.

REFERENCES

Bonet-Maury, P. (1953). *Bull. Soc. chim., Paris* 1066.
Ravenhill, J. R. and James, A. T. (1967). *J. Chromatogr.* **26**, 89.
Snyder, F. (1964). *Anal. Biochem.* **9**, 183.
Snyder, F. (1965). *Anal. Biochem.* **11**, 510.
Tittle, C. W. (1962). *Nuclear Chicago Technical Bulletin* No. 14.
Wang, C. H. and Willis, D. L. (1965). "Radiotracer Methodology in Biological Science", p. 187. Prentice-Hall Inc., Englewood Cliffs, New Jersey.
Wilde, P. F. (1965). "Thin Layer Chromatography", p. 29. United Trade Press Ltd., London.
Williams, R. R. and Smith, R. E. (1951). *Proc. Soc. exp. Biol. Med.* **77**, 169.

Chapter 10

Bidimensional Radiochromatogram Scanning

A. E. LOWE

*Biochemistry Department, Imperial College,
University of London, London, England*

The technique of bidimensional radiochromatography involves the preparation of chromatograms as with non-radioactive material. The resulting radioactive spots must then be detected and measured by some suitable method.

One quantitative method is to divide the chromatogram into an imaginary matrix of contiguous, equal areas and assay each of these areas separately with a pair of Geiger-Müller counters (one on each side of the paper). This method is particularly suitable for the soft β-emitters commonly used in biochemical studies. The movement of the G.M. counters with respect to the paper is done automatically and is usually referred to as scanning.

It will readily be appreciated that the requirements for statistical accuracy and/or spatial resolution on the one hand, and speed of scanning on the other are in opposition, and the difficulty of reaching an acceptable compromise between these opposed requirements is the inherent disadvantage of the method. To take a practical case: the division of the chromatogram into less than about 900 areas (that is a matrix of 30×30) gives an inadequate resolution for most purposes. To record a statistically significant number of counts per area it is not practical to count each area for less than an average time of 1–1·5 min. This gives a total *minimum* time for a total scan of 900 min or 15 hr. A demand for better resolution and/or statistical accuracy increases the total scanning time.

As a single experiment may result in 20 chromatograms, it is obvious therefore that a single scanner is of little use for the routine evaluation of bidimensional radio chromatograms. The installation at Imperial College consists of an assembly of eight scanners which have a common printing system recording the output data on punch tape for subsequent computer processing.

This system is a development of the scanners designed at the Istituto Superiore di Sanita in Rome (Frank *et al.*, 1959). In these machines the chromatogram is Sellotaped to a similar-sized piece of ordinary paper, the latter being fed into the platen of an electric typewriter fitted with a support behind the carriage so that the chromatogram hangs vertically behind, and clear of, the typewriter.

A pair of G.M. counters is fixed so that the chromatogram hangs between them. These counters feed an autotimed scaler which, at the conclusion of the pre-set counting period, transmits the accumulated count to the relevant number keys of the typewriter. A four-figure count is recorded, the non-significant zero's being replaced by spaces. As the printing occurs the normal typewriter action moves the carriage, and hence the chromatogram by an equal amount in relation to the fixed G.M. counters. The typewriter thus simultaneously records the activity and moves the chromatogram into the next position to be counted.

This, however, leaves the calculation of "spot" activities (a spot covers a number of squares on the matrix) to the operator. Moreover it is not possible to change the size of the unit counting area as this is fixed by the typewriter spacing.

It was required to provide a system which would (a) calculate "spot" totals automatically; (b) enable the unit counting area to be readily changed; (c) have the possibility of counting thin-layer plates without too much modification.

The computerized version was designed to meet these requirements.

The typewriter is no longer used and instead the relative movement between the chromatogram and G.M. counters is obtained by a motor-driven mechanical system.

A frame running in nylon slides, from which the chromatogram is suspended, provides for the vertical movement of the chromatogram and the G.M. counters are fixed to saddles which move horizontally across the chromatogram.

Electric motors drive these components via a system of gears and lead screws.

The scanning is done by moving the G.M. counters across the chromatogram in discrete steps in successive horizontal lines. In each position an autotimed scaler registers the radioactivity for a pre-set time. At the conclusion of each scanned line the G.M. counters return rapidly to the beginning of the next line.

The positioning of the G.M. counters and the frame is achieved by switching the electric motors as required. Each motor is controlled by a light-photocell system "inspecting" a plate drilled with a line of equidistant holes corresponding to the desired stopping positions. The light-photocell assembly is attached to the component whose position it is desired to control and travels along the plate, which is fixed.

Further photocells derive signals from other perforations on the plates to: (a) control the reversal of the horizontal motor after each line; (b) stop the machine at the completion of the scanning; and (c) send information to the print-out system regarding the coordinate position of each scanned area. The multiple light-photocell systems are mounted in two holders, one of which is attached to the G.M. counters and the other to the frame.

In order to conserve total scanning time two pre-set times for the timed counting period are available. The shortest of these is a minimum time allowed for the counting of a single area. If, in this time, the scaler has accumulated a significant number of counts over background, the counting period is extended by a multiplier. The basic time, threshold and multiplying factor are under the control of the operator by means of switches.

When the count is below the threshold value no record is made on the punched tape.

A radioactive standard is fitted in such a position that it is counted at the beginning of each line of the scan. From the radioactive standard readings the computer can calculate the efficiency of the G.M. counter and can also check that the standard deviation of the set of readings indicates proper functioning of the machine.

At the conclusion of each overthreshold counting period the shared tape punch records the following data—the number of the machine from which the data originates, the total count, the time taken for the count, and the coordinate position of the area counted.

The machines are arranged on a time-sharing system so that no two machines require to punch data simultaneously.

The punched tape is sent to the Atlas computer for processing with a suitable programme tape. When processed the data is presented in the form of a "map" in which the actual count per unit time (less background) for each area is printed in the appropriate position on a piece of paper. The background is fairly constant and its value is supplied to the computer on the programme tape. Beneath the map a series of spot totals appears. Each total is positioned so that it is immediately below the spot to which it refers. These are again in counts/unit time less background. Beneath these again appear another set of spot totals corrected for G.M. counter efficiency. These totals may either be in disintegrations per min. or in micro micro curies. Where two or more spots are not fully separated the correct activity is allocated to each spot.

The G.M. counters are Tracerlab gas flow counters as used on their strip scanner, but run on 98% argon 2% isobutane instead of helium/isobutane. They are however quite reliable with this mixture and G.M. counter failure is rare. As it is necessary to supply eight pairs of counters the gas is mixed automatically at the time of use and this reduces the cost as against ready-mixed gas by at least 50%.

A cover plate is pressed over the front of the counter which has a window aperture equal to the desired unit counting area. This is covered with Mylar film ·015 in thick.

Overall efficiency is about 11%; linearity and reproducibility is $\pm 5\%$, measurements of the above being made by "spotting" known activities on to Whatman No. 1 chromatography paper and recovering the counts from the computed spot totals (the weakest spots being ten times background and the range of spot activities not greater than 15 : 1).

One of the drawbacks of the computerized system is that a high proportion of the notional cost of the processing goes in the layout of the "map". As the non-computerized version of the scanner already does this adequately the computation is apparently very costly. If the computer was required to produce a list of spot totals only, the "map" being produced by other means, processing costs would fall by about 75%.

A multiple automatic system of this sort is only viable if all the component parts are extremely reliable. The mechanical parts are made to good engineering practice standards. The electronic circuits are designed around Mullard 100 kc/s circuit blocks on plug-in printed circuit boards. Very few faults occur in either the mechanical parts or electronic circuits. The chief cause of breakdown has been the unreliability of the lamps used for illuminating the photocells. The system is designed in such a way that a failure in one scanner only very exceptionally affects the remainder of the system.

Machines are in existence with unit counting areas of side 0·4 in. and 0·2 in. Changing from one size to another is simply done by substituting one set of perforated plates and G.M. counter windows for another.

Although the design of the scanners allows for the counting of T.L.C. plates by laying the machine on its side, and removing one G.M. counter, this has not yet been required.

The system has been in almost continuous use on a 24-hour day basis at Imperial College for more than two years, the number of machines increasing from four to eight during the first year of working.

[A more detailed description of the scanning system will be published elsewhere later.]

REFERENCE

Frank, M., Chain, E. B., Pocchiari, F., Rossi, C., Ugolini, F. and Ugolini, G. (1959). Selected Scientific Papers, Instituto Superiore di Sanita, II, (i), 75.

Chapter 11

The Evaluation of Radiochromatograms using a Spark Chamber

B. R. PULLAN

Department of Medical Physics, University of Manchester, Manchester, England

Spark chamber systems are being developed to increase the speed with which distributions of activity on radiochromatograms can be visualized. A spark chamber is suitable for this purpose because it is a radiation detector which is able to produce a visible spark in the position at which charged particles pass through it. Essentially the detector consists of two parallel electrodes placed in a suitable gas and between which an electric field can be created. A charged particle such as a β-particle passing through this high field will leave a trail of ions in the gas. The ions will be accelerated towards the electrodes and under the correct conditions can lead to a visible spark, the spark being at the position at which the charged particle passed through. If a spark chamber is made in such a way that one of the electrodes is transparent to β-particles and a chromatography medium which contains spots of β-ray emitting radioactive isotopes is placed against this electrode, sparks will be formed over the spots. Photographing the sparks with an extended exposure produces a picture similar to an autoradiograph, Fig. 1, but with an exposure 1000 or more times shorter than would have been necessary with conventional autoradiography using photographic film.

A. Spark Chamber Designs

Two types of chamber have been built and used for the evaluation of radiochromatograms. The first to be developed, Pullan *et al.* (1966), was a crossed wire chamber.

1. CROSSED WIRE CHAMBER

The basic geometry of the electrode structure is shown in Fig. 2. The two electrodes consist of arrays of parallel wires, the wires in the two electrode planes being mutually perpendicular and separated by 0·1 in. In a practical version of this type of chamber the cathode or earth electrode consists of 24-gauge tinned copper wires stretched over a perspex frame as shown in Fig. 3 and with a spacing between adjacent wires of 0·125 in. The anode or positive electrode is constructed using 0·004 in. diameter stainless steel wires also stretched over a perspex frame with a spacing between adjacent wires of 0·125 in. The wires are individually tensioned by means of a nut and bolt

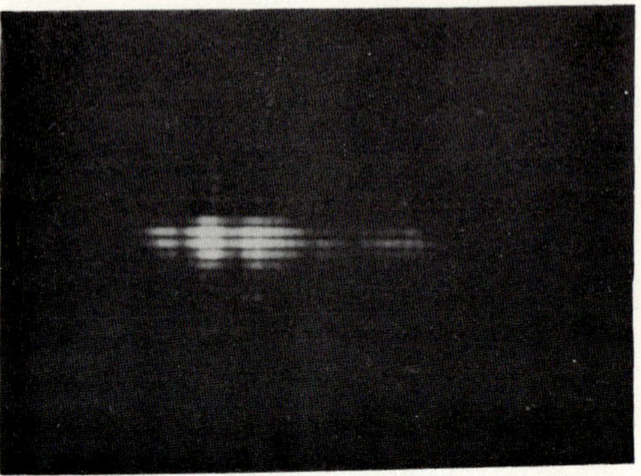

FIG. 1. A spark chamber autoradiograph of a ^{14}C radiochromatogram, exposure time 10 min.

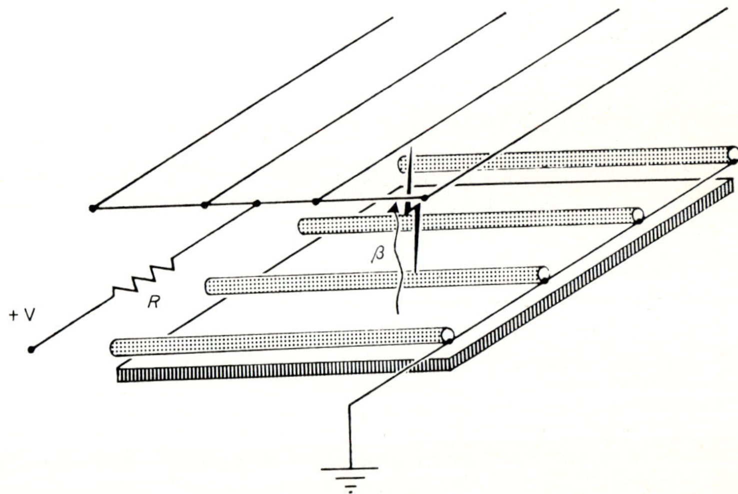

FIG. 2. Schematic diagram of the electrode arrangement of a crossed wire spark chamber. A chromatography plate is shown under the cathode wires and a β-particle travelling from it produces a spark between the cathode and anode.

at one end of each wire and are bent over bars near to the ends of the frames; this is so that when the electrodes are put together the separation between the wires increased at the edges of the chamber. It was found necessary to separate the electrodes at the edges in order to stop spurious sparking at the wire ends. The area between the two parallel sections of the electrodes is that which is sensitive to radiation emanating from a chromatography medium placed under and close to the cathode wires. In the version shown in Fig. 3 the chromatography plate is placed on the removable section in the cathode electrode frame.

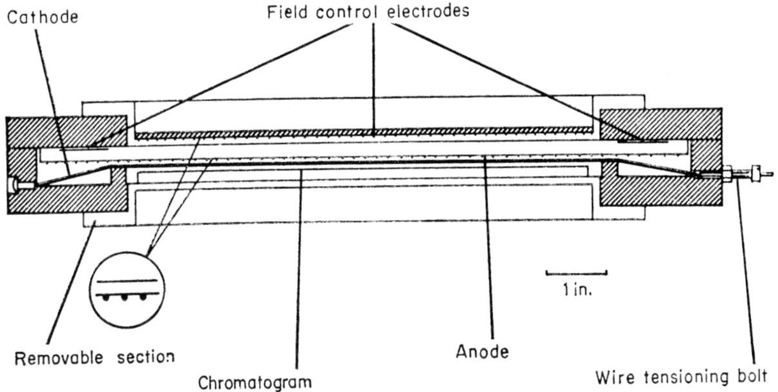

FIG. 3. Cross section view of a practical version of a crossed wire spark chamber showing a chromatography plate placed in position on a removable section.

In early versions of this chamber difficulties were experienced with spurious sparking in the regions where the electrode planes separated and it was found necessary to add electrodes to control the electric field at the edges of the chamber. These are shown in Fig. 3. One field control electrode is placed over the sensitive region and connected to the earth wires, a second electrode surrounds this and is connected to the positive electrode or anode.

The use of perspex in the construction of the electrode frames has proved a source of trouble because of its poor structural stability and more recent versions have been made using metal and C.P.3 plastic produced by Formica Ltd. Difficulties have also been experienced with this design in maintaining sufficient accuracy of construction to obtain uniform response. These difficulties together with a desire to find a quicker and easier method of making chambers lead to a spiral cathode version in which high structural accuracy can be maintained whilst employing a very simple method of construction.

2. SPIRAL CATHODE CHAMBER

The basic geometry of this construction is shown in Fig. 4. The chamber consists of a number of individual elements, each element having a spiral of wire constituing the cathode, the anode being a fine wire running along the axis of each spiral. A drawing of a practical version is shown in Fig. 5. The cathode is made of 24-gauge tinned copper wire and the anode is 0·004 in.

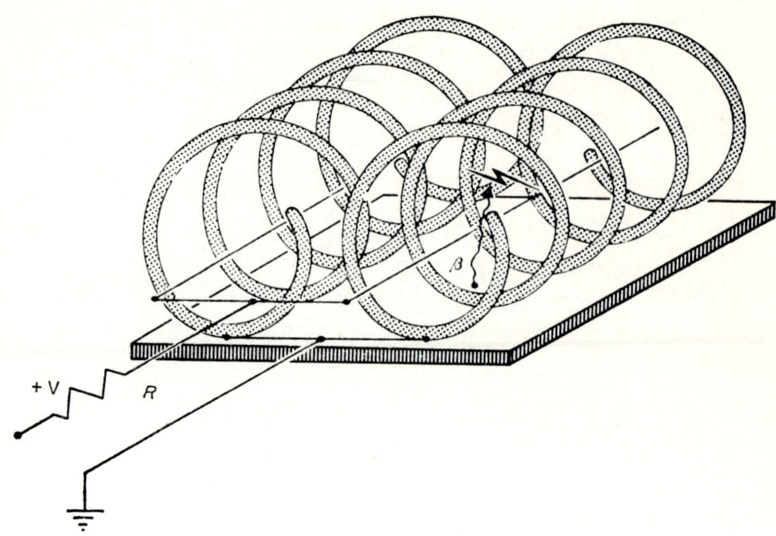

Fig. 4. Schematic diagram of the electrode arrangement of a spiral cathode spark chamber. A chromatography plate is shown in position under the spiral cathodes and a β-particle travelling from it produces a spark between the cathode and anode.

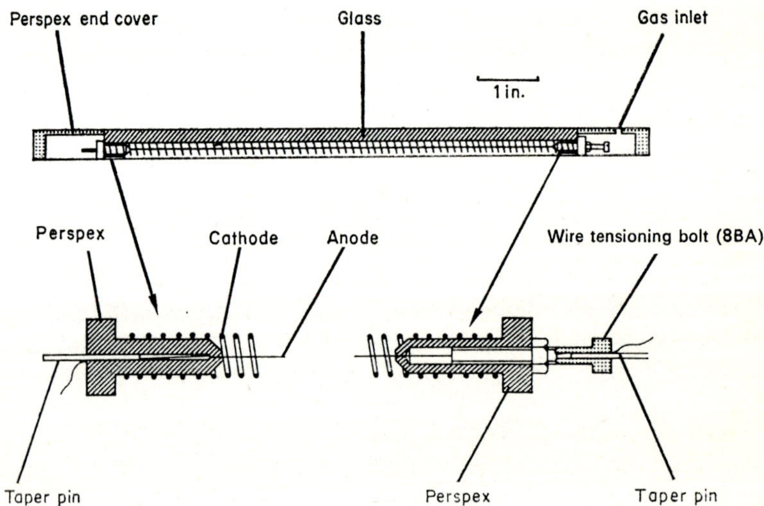

Fig. 5. Cross section view of a spiral cathode spark chamber with detailed views of the method of supporting and tensioning the anode wires.

diameter stainless steel wire. The cathode spirals are wound on a mandrel so that their outside diameter is 0·187 in. and their pitch 0·063 in. The winding operation can be done quickly and accurately using a screw cutting lathe and guiding the wire onto the mandrel using an attachment in the lathe tool post. After winding the spirals are washed in carbon tetrachloride and glued to a glass plate using Araldite adhesive type AV100. The araldite is spread in a thin uniform layer over a glass plate using a nylon roller and the spirals are pressed into the glue by a jig which holds them accurately in place during the glueing operation. The jig consists of a plate with accurately milled parallel slots into which the spirals are pressed. The jig, spirals and glass plate are put together and baked for 15 min at 90°C, after which the jig can be removed leaving the spirals stuck to the glass. Electrical connection is made between the elements by soldering a wire across the ends of the spirals. A frame is glued around the glass plate of which the perspex end covers (Fig. 5) form part, also strips of perspex 0·375 in. × 0·031 in. are glued across both ends of the spirals and fixed to the frame. This is necessary to strengthen the end of the spirals. A gas inlet is put in one of the end covers and gas flows into the region of the spirals through the spaces between the perspex plugs.

The chromatogram to be evaluated is placed in a well in a flat sheet of C.P.3 plastic so that its active surface is about 1/64 in. above the top of the well. The spark chamber is then placed on top of the chromatogram.

B. Operation

1. CROSSED WIRE CHAMBER

After the chromatography plate has been placed in a chamber with its active surface close to the cathode wires counter gas is flowed through. Two gas mixtures have been used successfully; these are 90% argon, 10% methane and argon methylale mixtures. The argon methane mixture can be bought ready mixed in cylinders and argon methylale mixture can be made by bubbling commercial argon through methylale at 0°C and passing this resulting mixture directly into the chamber. If the second gas mixture is used, gas escaping from the chamber should be removed from the working environment. After opening it is necessary to flush a chamber for at least 10 min before uniform sensitivity can be obtained and the flow of gas should be maintained during the whole of an exposure.

The electrical circuit for operating a chamber is very simple and is shown schematically in Figs 2 and 4. The positive terminal of a high voltage supply is connected through a resistance R, Figs 2 and 4, to the anode of the chamber; the cathode is connected to the earth terminal. Each spark discharges the capacity of the chamber and the voltage then recovers on the time constant of the series resistance R and the capacitance of the chamber. With a capacity of 150 pF the minimum resistance for stable operation is of the order of 10 MΩ. This gives a dead time of about 5 msec and stable operation for approximately 200 V above the spark threshold, i.e. the chamber operates from 3300 to 3500 V. With higher values of series resistance longer operating regions can be achieved and for a resistance of 250 MΩ it is possible to operate for approximately 1000 V above the spark threshold. A value between 10 and 80 MΩ is usually suitable and it is useful to be able to vary the value within this range.

The power supply used should operate up to about 4500 V and be stabilized to better than 1%.

The most convenient method of recording the distribution of sparks and therefore the distribution of activity is to photograph the chamber with an extended exposure using a poleroid camera set at about f64 and 3000 A.S.A. poleroid film. This must be done either in a darkened room or in a light proof box. The length of exposure required varies between about 10 sec and 1 hr depending upon the activity of the chromatogram.

A convenient method of comparing a spark chamber autoradiograph on poleroid film with a chromatography plate is to project it back onto the plate using an episcope.

When a newly-constructed chamber is first tried it is not unusual to experience trouble with repeated sparking or instability at a number of points, also the sensitivity to radiation may be poor. The stability can usually be improved by making sure that there are no sagging wires and by replacing bad wires. The longer a chamber is operated the more stable it tends to become and it is advisable to let a new chamber spark vigorously for a few hours before attempting to use it. Washing the electrodes in a solution of iodine in alcohol also helps to improve stability. Low sensitivity to radiation is often due to field distortions caused by charge picked up on the perspex surface over the anode wires and an antistatic preparation should be applied to this surface. One of the difficulties in operating this type of chamber with thin-layer chromatograms is that loose powder is attracted from the plate and deposited on the electrodes and the surface over the anode. This can lead to contamination which is difficult to remove because this kind of chamber is not easy to clean.

2. SPIRAL CATHODE CHAMBER

This type of chamber is also operated with a continuous flow of gas but in this case only 90% argon, 10% methane gas mixture has been used successfully. It is only necessary with this design to flush the chamber for 30 sec before uniform response is obtained.

The electrical circuit requirements are similar to those of the crossed wire chamber and the design described becomes sensitive at 1800 V and goes unstable at approximately 2100 V even for high values of series resistance.

The power supply should be stable to better than 0.5% and it is usual to operate with a series resistance of between 20 and 40 MΩ.

The chamber can be photographed in the same way as before except that an aperture of about $f32$ is necessary when using 3000 A.S.A. poleroid film.

This type of chamber is much less temperamental than the crossed wire version and the four that have been built so far have worked immediately with no troubles at all. These chambers are very rugged and can be scrubbed with a nail brush if decontamination becomes necessary. The problem of thin layers being pulled off by the electric field does not occur because the electric field does not extend beyond the cathode as much as it does with the crossed wire design.

C. Performance

1. AREA OF SENSITIVITY

Crossed wire spark chambers have been built with sensitive areas up to 8 in. × 8 in. It is thought, however, that increasing difficulties would be experienced if much bigger chambers were attempted using this method of construction. A spiral cathode chamber has been made with a sensitive area 12 in. × 19 in. and the method of construction described is easily adaptable to any size.

2. UNIFORMITY OF RESPONSE

Figure 6 shows spark chamber autoradiographs of radioactive resolution test charts; (a) and (b) were taken with a spiral cathode chamber and (c) and (d) with a crossed wire chamber. The uniformity of response can be seen to be far superior with the spiral cathode type. In Fig. 6(c) and (d) the anode wires run across the picture and it can be seen that good uniformity of response is maintained along the anode wires but is poor in a direction at right angles to this.

3. RESOLUTION

The resolution charts of which spark chamber autoradiographs are shown in Fig. 6(a) and (c) were painted on paper with [^{14}C]labelled ink and (b) and (d) with [^{3}H]labelled ink. No estimate of the surface activity of these targets has been made. The targets consist of nine $10°$ wide segments of activity separated by $10°$ wide non-active segments, each segment being 7 cm long. The autoradiographs were taken with the test charts 1 mm from the cathode in all cases. The results show that the spiral and crossed wire chambers have approximately the same resolution for segments which are perpendicular to the spirals but as would be expected from the chamber structure this falls off relative to the crossed wire type for segments parallel to the spirals. It is interesting that if ^{3}H activity is confined to one side of the axis of a

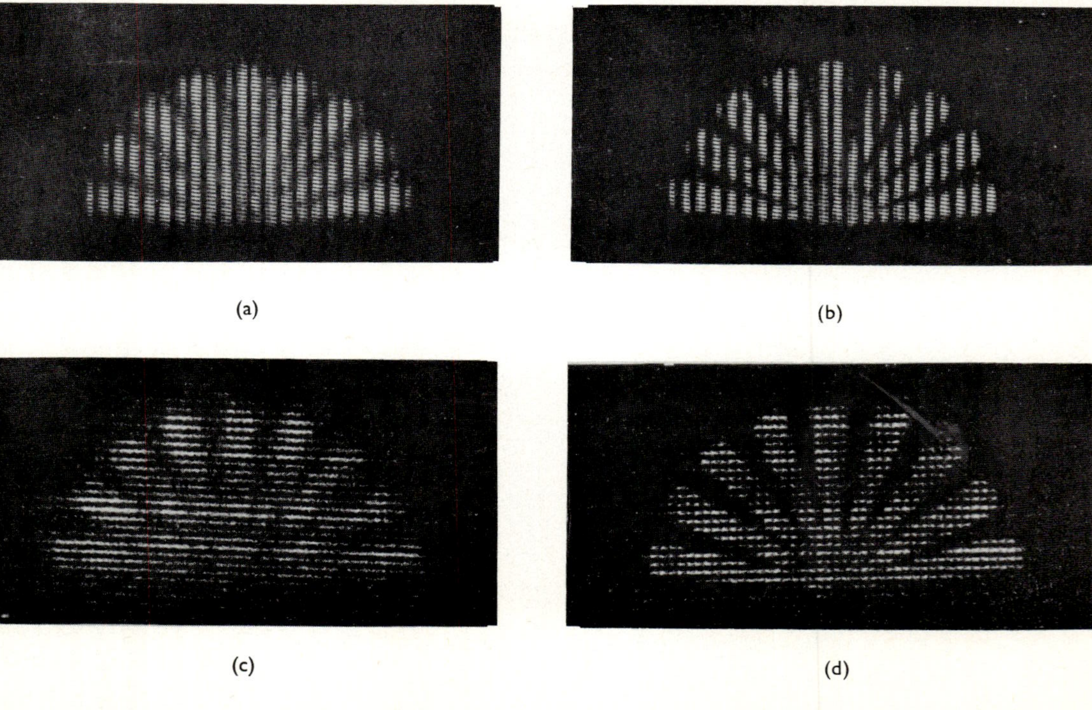

Fig. 6. Spark chamber autoradiographs of radioactive resolution test charts (a) ^{14}C test chart and (b) ^{3}H test chart both taken using a spiral cathode chamber, (c) ^{14}C test chart and (d) ^{3}H test chart both taken using a crossed wire spark chamber.

spiral the sparks form between the anode and the cathode on that side. This gives this type of chamber a resolution in a direction at right angles to the spirals which is greater than was expected.

The results indicate that the resolution of the crossed wire type of chamber for ^{3}H is about 2 mm and for ^{14}C about 4 mm. For the spiral cathode chamber the corresponding figures are 3 mm and 6 mm when the resolution is averaged over all directions. The resolution depends upon the separation between chromatography plates and the cathode and it is desirable to keep this separation less than 1 mm.

4. SENSITIVITY

It is difficult to give a meaningful figure for the absolute efficiency for detecting β-particles as this depends critically on a multitude of factors. It is felt that a more useful figure is the minimum detectable activity in a spot of a given diameter on a particular medium and with a fixed exposure. The results given are for 0·5 cm diameter spots on Watmans No. 1 chromatography paper and with an exposure for 10 min. Figure 7(c) shows a spark chamber autoradiograph taken with a spiral cathode chamber of seven spots of ^{14}C of activity 1·9, 0·96, 0·48, 0·24, 0·12, 0·06 and 0·03 mμ C, i.e. approximately 4500 d.p.m. down to 70 d.p.m. 0·03 mμ C is regarded as the lowest activity which can be usefully visualized under these conditions. Figure 7(a) and (b) show the same test object taken with a crossed wire chamber. In (b) 90% argon 10% methane gas mixture was used and in (a) argon methylale mixture was used. Both types of chamber are equally sensitive to ^{14}C.

It is apparent from these results that the resolution depends upon the exposure and that at the exposure necessary to visualize the weak spots the more active spots are spread out considerably.

In the case of all these pictures the rate of sparking was limited by the dead time. This may also be the case if no radioactive object is placed in the chamber, the background sparking rate being limited by the dead time. The sparks will be divided between the active spots and the background in approximate proportion to their activity. This means that a weak spot will tend to be suppressed in the presence of a strong one as also will the background. Because of this spots differing in activity by more than 100 times are difficult to visualize in a single exposure. It is advisable when dealing with a wide range of activities to do a short exposure to show the most active spots with good resolution. These should then be cut out or a small absorbing filter placed over them and another exposure made to show the weak spots.

The sensitivity to ^{3}H under the conditions specified is considerably less than for ^{14}C. For the crossed wire chamber the lowest activity spot which can be visualized is of the order of 0·3 mμ C and for the spiral chamber 3 mμ C. These figures are very approximate as the detectability depends

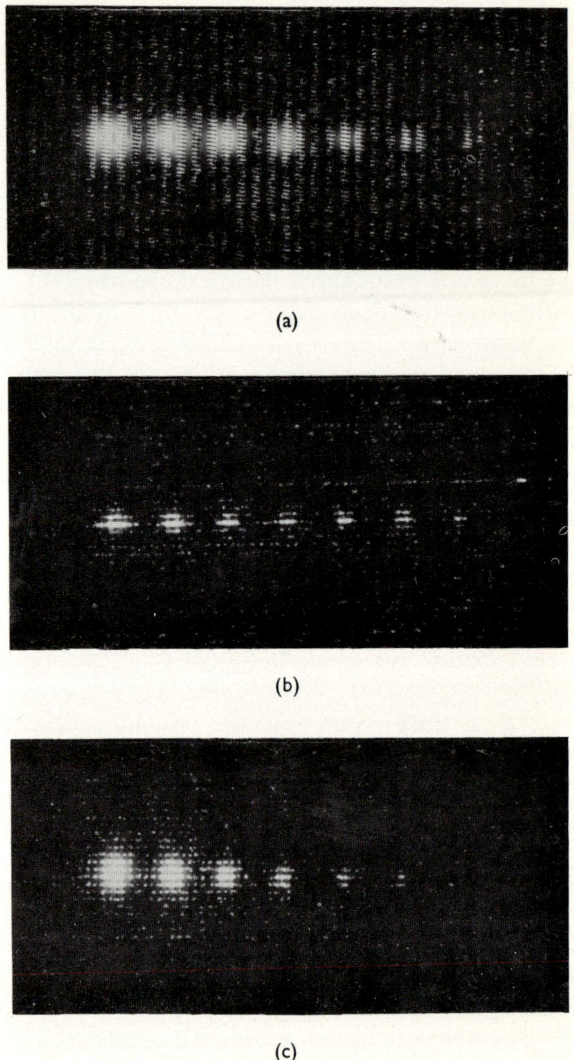

(a)

(b)

(c)

Fig. 7. Spark chamber autoradiographs of seven spots of activity which are from left to right, 1·9, 0·96, 0·48, 0·24, 0·12, 0·06 and 0·03 mμ C. These spots were 0·5 cm diameter and on Watmans No. 1 chromatography paper. The exposure in each case was 10 min.

critically on the volume distribution of activity through the chromatography medium and this is very non-reproducible.

The difference in sensitivity to ^{3}H for the two types of chamber is probably due to differences in electric field geometry in the two types. In the crossed wire type the electric field extends beyond the cathode more than in the spiral type and ions created in the region between the chromatography plate and the cathode can be dragged into the sensitive region.

D. Quantitation

With a crossed wire chamber it is relatively easy to extract electrical signals corresponding to spark positions and to process these by either digital or analogue methods (Pullan *et al.*, 1966). Electronic systems for this purpose have been built and used but showed that the uniformity of response which was being achieved with even the best of the crossed wire chambers was insufficient to warrant the extra complexity of the equipment. It is the opinion of the author that the reproducibility of beta counting direct from chromatography plates for ^{14}C and ^{3}H does not warrant the effort of extracting quantitative data direct from a spark chamber. It is felt that the proper role of the spark chamber is in the rapid determination of the positions of active spots on radiochromatograms so that they can be cut out and counted by a more accurate method such as liquid scintillation counting.

E. Conclusion

The spark chamber as an instrument for the evaluation of radiochromatograms is still in an early stage of development. The spiral cathode spark chamber, however, provides an easily used and reliable means of making rapid qualitative autoradiographs within certain limitations. These are that if it is required to resolve two adjacent spots of activity their edges must be separated by more than 6 mm in the case of ^{14}C and 3 mm in the case of ^{3}H and that spots of activity do not differ by more than 100 times if they are to be visualized in a single exposure. Within these limitations the spark chamber works well.

Acknowledgement

I wish to express my gratitude to Mr S. Carlile for his skill and enthusiasm in helping me with this work.

REFERENCE

Pullan, B. R., Howard, R. and Perry, B. J. (1966). *Nucleonics* **24**, 72.

Author Index

The numbers in *italics* indicate the pages on which names
are mentioned in the reference lists

Subject Index